Probabilistic Optimisation
of Composite Structures

Machine Learning for Design Optimisation

Computational and Experimental Methods in Structures

ISSN: 2044-9283

Series Editor: Ferri M. H. Aliabadi *(Imperial College London, UK)*

This series will include books on state-of-the-art developments in computational and experimental methods in structures, and as such it will comprise several volumes covering the latest developments. Each volume will consist of a single-authored work or several chapters written by the leading researchers in the field. The aim will be to provide the fundamental concepts of experimental and computational methods as well as their relevance to real-world problems.

The scope of the series covers the entire spectrum of structures in engineering. As such it will cover both classical topics in mechanics, as well as emerging scientific and engineering disciplines, such as: smart structures, nanoscience and nanotechnology; NEMS and MEMS; micro- and nano-device modelling; functional and smart material systems.

Published:

Vol. 15 *Probabilistic Optimisation of Composite Structures:*
Machine Learning for Design Optimisation
by Kwangkyu Alex Yoo (Imperial College London, UK &
Deep.Meta, UK), Omar Bacarreza (Imperial College London, UK &
ORCA Computing, UK) and M. H. Ferri Aliabadi
(Imperial College London, UK)

Vol. 14 *Uncertainty Quantification of Guided Wave Structural Health*
Monitoring for Aeronautical Composite Structures
by Nan Yue (Delft University of Technology, The Netherlands),
Zahra Sharif Khodaei (Imperial College London, UK) and
M. H. Ferri Aliabadi (Imperial College London, UK)

Vol. 13 *Mathematical Methods and Models in Composites*
(Second Edition)
edited by V. Mantič (University of Seville, Spain)

Vol. 12 *Wear in Advanced Engineering Applications and Materials*
edited by Luis Rodríguez-Tembleque (Universidad de Sevilla, Spain),
Jesús Vázquez (Universidad de Sevilla, Spain) and
M. H. Ferri Aliabadi (Imperial College London, UK)

More information on this series can also be found at http://www.worldscientific.com/series/cems

(Continued at end of book)

Computational and Experimental Methods in Structures – Vol. 15

Probabilistic Optimisation of Composite Structures

Machine Learning for Design Optimisation

Kwangkyu Alex Yoo
Imperial College London, UK & Deep.Meta, UK

Omar Bacarreza
Imperial College London, UK & ORCA Computing, UK

M H Ferri Aliabadi
Imperial College London, UK

World Scientific

NEW JERSEY · LONDON · SINGAPORE · BEIJING · SHANGHAI · TAIPEI · CHENNAI

Published by

World Scientific Publishing Europe Ltd.

57 Shelton Street, Covent Garden, London WC2H 9HE

Head office: 5 Toh Tuck Link, Singapore 596224

USA office: 27 Warren Street, Suite 401-402, Hackensack, NJ 07601

Library of Congress Cataloging-in-Publication Data

Names: Yoo, Kwangkyu Alex, author. | Bacarreza, Omar, author. | Aliabadi, M. H. author.
Title: Probabilistic optimisation of composite structures : machine learning for design optimisation / Kwangkyu Alex Yoo, Omar Bacarreza, M H Ferri Aliabadi.
Description: New Jersey : World Scientific, [2025] | Series: Computational and experimental methods in structures, 2044-9283 ; vol. 15 | Includes bibliographical references.
Identifiers: LCCN 2025000009 | ISBN 9781800616844 (hardcover) |
    ISBN 9781800616851 (ebook for insitutions) | ISBN 9781800616868 (ebook for individuals)
Subjects: LCSH: Structural optimization--Mathematical models | Structural optimization--Data processing. | Composite construction--Mathematical models. | Composite construction--Data processing. | Machine learning--Industrial applications.
Classification: LCC TA658.8 .Y66 2025 | DDC 624.1/77130285--dc23/eng/20250228
LC record available at https://lccn.loc.gov/2025000009

British Library Cataloguing-in-Publication Data
A catalogue record for this book is available from the British Library.

For any available supplementary material, please visit
https://www.worldscientific.com/worldscibooks/10.1142/Q0496#t=suppl

Desk Editors: Nambirajan Karuppiah/Gabriel Rawlinson/Shi Ying Koe

Typeset by Stallion Press
Email: enquiries@stallionpress.com

Preface

This monograph presents multi-fidelity probabilistic optimisation to address the computational challenges inherent in the stochastic design philosophies applied to aircraft composite structures. In aircraft design using composites, uncertainties in factors such as geometry and material properties can lead to increased weight and degraded performance due to conservative safety factors. This, in turn, results in higher carbon emissions, wasted materials, and increased maintenance costs. To tackle these challenges, extensive research has been conducted to enhance computational efficiency and address these uncertainties in the early design stages. However, conventional approaches employing a single high-fidelity model (HFM) have not adequately tackled the significant computational challenges posed by large-scale and nonlinear composite structures.

The innovative multi-fidelity formulations developed herein blend HFMs and low-fidelity models (LFMs), significantly enhancing computational efficiency through the use of machine learning techniques, such as artificial neural networks and nonlinear autoregressive Gaussian processes. To further improve efficiency over conventional probabilistic optimisation methods, a multi-level optimisation approach and a new sampling strategy are introduced into multi-fidelity formulations for the first time.

Several engineering examples, including aircraft mono-stringer composite structures, demonstrate the accuracy and efficiency of the developed methods when applied to various reliability and robustness analysis techniques, such as Monte Carlo simulation, the first-order reliability method, and the second-order reliability method. These examples, which involve mechanical and thermo-mechanical loads, highlight the broad range of potential applications of the new approach developed in this monograph.

About the Authors

Kwangkyu Alex Yoo is a senior machine learning scientist at Deep.Meta in London, a Google-funded AI start-up. Alex completed his PhD in multi-fidelity probabilistic optimisation for composite structures at Imperial College London in 2021. Following his PhD, Alex served as a research associate in industrial machine learning at the University of Cambridge. Prior to his doctoral studies, he was a research engineer specialising in offshore engineering at Hanwha Ocean in Korea. Alex has consistently developed machine learning and optimisation algorithms to tackle technical challenges across design, manufacturing, operations, and supply chains. His expertise lies in machine learning, multi-fidelity modelling, and optimisation under uncertainty, making him a versatile contributor to both academic and industrial advancements. His work is widely recognised through journal articles, conference papers, patents, invited talks, and commercial projects.

Omar Bacarreza is a senior machine learning scientist at ORCA Computing, working on quantum computing applied to machine learning and optimisation. He has a PhD in computational mechanics from the Czech Technical University in Prague. Dr Bacarreza has more than 30 publications on optimisation, composite materials, and structural health monitoring and has participated in several projects involving academic and industrial partners. He has used high-performance computing in computational mechanics and machine-learning-assisted stochastic design and optimisation of aircraft components. He also has experience with AutoML techniques for improving and finding better deep learning models.

M H Ferri Aliabadi is a professor of aerostructures at Imperial College London. He has worked in the field of solids and structures and has established an international reputation for his achievements in the development of computational methods related to fracture and damage mechanics. Prof. Aliabadi has pioneered a new generation of boundary element methods and is noted for his contributions to other fields. During the past decade, he has pursued research and development in structural health monitoring (SHM) for composite airframes in collaboration with the aeronautics industry (Airbus and Leonardo), supported by significant EU funding. He has been Principal Investigator (PI) and coordinator of several CleanSky EU projects, including SMASH and SCOPE, and has coordinated the SHM platform for the wing in the SARISTU project. His latest project is CleanSky II, a core partnership with SHERLOC, in which he is the PI and coordinator, seeking to develop the next generation of smart (highly sensorised) composite airframes. Prof. Aliabadi has published 500 papers in leading international journals and 65 books related to experimental and computational methods in solids and structures.

Contents

Chapter 1

Introduction

Innovative and sustainable design technologies are associated with improvements in energy efficiency and reductions in carbon emissions, due in part to the increased use of composite materials and the fundamental advantages that they offer, specifically their high strength and light weight [1]. Structures using these materials are widely employed in a vast range of infrastructure manufacturing industries, such as aircraft, renewable energy, naval architecture, and automobiles. However, traditional design approaches do not consider design uncertainties that every engineering system possesses throughout its entire lifecycle. These uncertainties originate from design, manufacturing, operation, and ageing, which can deteriorate the performance quality of a structure. A design process that does not consider these uncertainties may also lead to either premature structural failure or highly conservative designs relying on high safety factors. Hence, it is desirable to define a probability of success and sensitivity to variations, which are referred to as reliability and robustness, respectively. The probability in this context, also known as the probability of failure, refers to a scenario where the design fails to satisfy particular criteria. Failure does not necessarily represent a catastrophic structural collapse but rather suggests that the structure is unable to offer the target structural performance. Reliability analysis estimates the probability of failure for each structural element. In contrast, robustness pertains to the stability of the performance quality of the designed structure in response to variations in the manufacturing process or environmental conditions. These variations associated with manufacturing and operation lead to

1

additional costs throughout the lifecycle, including unscheduled inspections, repairs, and maintenance [2]. The process of estimating robustness is named robustness analysis [3]. These reliability and robustness analyses provide many benefits to engineers during the early stages of the design process. Reliability analysis allows engineers to comprehend how different engineering parameters affect the reliability of structures they have designed and to establish a design philosophy aimed at improving the structures' overall probability of success. Robustness analysis enables engineers to account for possible variations influencing the performance quality and obtain a more stable design without removing or minimising their sources [4].

Even though reliability and robustness are significant for every structure, they are of essential importance in the infrastructure manufacturing industry for the purpose of improving energy efficiency by reducing structural weight. This can tackle the inefficiencies in traditional design approaches caused by the use of large safety factors mandated by international standards and design specifications for each field. These factors generally deliver a conservative and deterministic design to fulfil the required safety levels. For instance, aircraft structures are required by the Federal Aviation Administration (FAA) to incorporate a safety factor of 1.5 for external loads on the structures while adding additional factors to enhance the safety levels concerning material properties, manufacturing effects, temperature effects, etc. [5].

Probabilistic design optimisation incorporates reliability and robustness analyses into its optimisation process, and it considers the statistical characteristics caused by the uncertainties associated with the design and manufacturing stages. Therefore, it enables the final design to deliver more reliable and robust engineering features to the industry, which seeks ways to preserve a specified safety level while achieving light weight and fuel efficiency, as well as net-zero emissions — an international goal ('Net Zero by 2050') announced by the International Energy Agency (IEA) [6]. Such an optimisation process accommodates different methods to carry out either reliability or robustness analysis depending on the objectives of optimisation. In general, Monte Carlo simulations (MCSs) are a typical statistical method used to estimate reliability and robustness. Reliability can also be predicted using non-statistical methods, including first-order and second-order reliability methods, or FORMs and SORMs, respectively [7]. These two methods require approximating a limit state function using the first-order and second-order Taylor series expansions, respectively. More details of these methods can be found in many books on reliability and robustness analyses [8, 9].

Reliability and robustness analyses require a massive number of experimental tests on the structure to investigate the effects of design uncertainties. However, these experimental tests are unfeasible due to high costs, long execution times, and the considerable trained workforce they require. Numerical methods, particularly the finite element method (FEM) for structural design, can be exploited to obtain solutions that are as accurate as experimental tests without necessitating substantial resources. Even if the FEM can provide a certain level of time savings, the estimation of reliability and robustness at every single design point remains challenging due to the high computational costs of conducting FEM simulations. Note that reliability and robustness analyses using MCS are expected to require tens of thousands of FEM simulations at each design point to reach a converged statistical result [8]. For example, if probabilistic optimisation using a genetic algorithm (GA) consists of 12 populations and 20 generations, while MCS for robustness analysis requires 10,000 simulations per population, the total number of FEM simulations amounts to 2,400,000. Even when a structural problem is simple enough for a single FEM simulation to take only a few seconds to execute, the total computational time for probabilistic optimisation can extend to nearly a month. As engineering problems in composite structures become more complex, involving large-scale structures with many design variables subjected to complicated physics, the time required for completing an FEM simulation for analysis will exceed several hours. This suggests that the total simulation time can take up to several months or even a few years depending on the problem features. To deal with this prohibitive computational challenge, many model approximation methods have been developed for use in place of the FEM model; such models are called surrogate models or metamodels [10–12]. These models enable engineers to carry out reliability and robustness analyses for precise probabilistic design optimisation with improved computational efficiency, cutting down FEM simulation times from a few hours to microseconds. Multi-fidelity modelling, introduced in computer science, is garnering ever-more attention for structural optimisation because it offers more significant computational gains than surrogate modelling while maintaining its accuracy [13]. Specifically, multi-fidelity modelling is expected to overcome complex and large-scale composite design problems whose computational costs make them highly prohibitive even for classical surrogate modelling.

1.1 Composite Structures in Aircraft Design

Composite structures are widely used in various manufacturing industries, including aircraft, wind turbine manufacturing, and automobiles. The recent developments in composite materials, including light weight and multifunctionality, provide many engineering advantages to the industry and have resulted in a dramatic increase in the usage of the materials for their products. With design approaches for environmentally friendly and sustainable development being incentivised to tackle global climate change, aircraft engineering is at the forefront of developing composite structures to introduce innovative aircraft design, along with achieving reduced fuel consumption through lightweight construction [14]. The primary parts using these composite structures in aircraft range from fundamental elements to large systems. It is not surprising that the Boeing 787 Dreamliner and the Airbus A350 XWB consist of approximately 50% and 53% composite structures, respectively. These percentages are on the rise, enabling even greater weight reduction in aircraft fuselages (commonly 20% lighter than aluminium) while achieving higher strength and extended lifespans [15].

One of the most appealing uses of composite structures is in designing the stiffened panel, which is an essential structural element in aircraft design. A stiffened panel allows a thin skin to support extreme loading in both tension and compression by using longitudinal stringers across the panel at a certain distance. The design of the stiffened composite panel should aim to maximise its strength, stiffness, and buckling load. It should be noted that the buckling load is not the maximum load that a structure can carry. This implies that the structure can hold several times the buckling load before structural failure, referred to as the nonlinear post-buckling strength. This post-buckling strength capability offers impressive potential for both weight and cost savings. Hence, designing a composite panel that will work under the buckling load all the time is a very conservative approach. A more lightweight composite structure can be designed if the extra strength under the post-buckling regime is put to use [4, 15–17]. Furthermore, thermo-mechanical loading has attracted significant attention in the aeronautics industry because of a massive surge in demand for high-speed and lightweight aircraft. This loading may cause untimely structural collapse when the aircraft is operated in extreme surroundings. In particular, the consideration of both thermal and mechanical loading can improve the design of composite aircraft since lightweight structures are typically susceptible to buckling under extreme conditions caused by high-speed operation [18–21].

A comprehensive coverage of the mechanics of composite structures can be found in a book by Jones [1]. In addition, overviews of nonlinear post-buckling and thermo-mechanical buckling, which are more specific to the work presented in this monograph, can be found in Chapters 5 and 6.

1.2 Probabilistic Design Optimisation

Industrial manufacturing processes of composite structures are inherently more complex than those of conventional metal and alloy structures, as they involve significant uncertainties in mechanical properties, geometry, and loading conditions associated with the entire lifecycle [22, 23]. One of the major challenges concerning traditional composite design approaches is that the final design often fails to adequately represent these uncertainties and instead presumes them through the use of large safety factors, hence resulting in conservative designs. These uncertainties assumed by the use of safety factors result in a designed structure that is inefficient, thereby increasing the requirements of raw materials for manufacturing and the fuel consumption for operation. The weight of the aircraft is particularly important to reduce carbon emissions from aviation while still meeting fundamental design requirements and capabilities. Hence, an innovative aircraft design approach should aim to minimise structural weight while providing a high safety level, achieved by considering the uncertainties associated with design and manufacturing processes.

An essential task would be to accurately estimate the statistical characteristics of structural performance influenced by uncertainties in engineering parameters, such as geometry, material properties, and external loading. For example, insufficient understanding of unpredictable features and errors in the computational model may introduce uncertainties at the early design stage. During the manufacturing stage of composite structures, a broad spectrum of flaws can lead to significant variations in material properties and geometry. However, traditional deterministic design optimisation methods do not make provisions for incorporating statistical measurements; consequently, this results in structures being over-optimised across their whole lifecycle. Even if these deterministic approaches find a global design that works quite well at the design point, it might show poor performance outside this point as well as a higher probability of failure. Thus, this design cannot guarantee reliability or robustness because it does not account for the significant influence of uncertainties [24, 25].

An approach to dealing with the above drawback is the probabilistic design optimisation methods, including reliability-based design optimisation (RBDO) and robust design optimisation (RDO) [16]. These methods are classified depending on their aims, rooted in differing philosophies. The purpose of RBDO is to minimise the probability of failure by accounting for design uncertainties so that the final design has a higher level of reliability using a probabilistic approach [15]. In comparison, RDO aims to reduce the variability in structural responses caused by unexpected deviations induced by the uncertainties [4]. A robust design developed using RDO achieves improved performance quality during the lifecycle when compared with a deterministic design. These two probabilistic design optimisation methods primarily incorporate reliability and robustness analyses into their processes. MCS can be exploited to estimate both reliability and robustness, while the first-order reliability method (FORM) and the second-order reliability method (SORM) are employed to predict the probability of failure. The procedure begins with searching the entire design space to obtain potential design points. Once obtained, reliability and robustness analyses are used to estimate the statistical characteristics of the design points related to the aim of probabilistic design. This is the primary feature that distinguishes the probabilistic design optimisation methods from traditional deterministic approaches. By exploiting the probabilistic philosophy during the optimisation process of aircraft structural elements using composite structures, engineers can understand the statistical nature influenced by the uncertainties associated with the design and manufacturing processes.

Although probabilistic design optimisation offers a reliable and robust structural design with improved efficiency over the whole lifecycle, it has not been implemented in the infrastructure design field due to its tremendous computational resource requirements [26]. A comprehensive overview and theoretical background of probabilistic design optimisation can be found in a book by Choi [8]. In addition, outlines of RBDO and RDO and a comparison between them highlighting the work presented in this monograph can be found in Chapter 2.

1.3 Surrogate Modelling

Surrogate modelling, also known as metamodelling, aims to approximate computationally expensive models based on the finite element method (FEM) for structural optimisation [10, 27]. This allows the FEM models to be represented by a computationally cheap alternative model. The alternative model enables probabilistic design optimisation to overcome the prohibitive computational challenge of evaluating the influence of design uncertainties. The computational time when using a surrogate model is considerably shorter than that of the original FEM model, usually ranging from hundreds to tens of thousands of times faster, while offering comparable accuracy. This facilitates significant improvements in the probabilistic design optimisation process by reducing the total optimisation time from several weeks to several hours while preserving a solution accuracy of less than 1% error. Such a surrogate model is created by collecting experimental observations or conducting numerical simulations. When sufficient information is collected, the surrogate modelling method creates a surrogate model by training and testing the collected data. This modelling process might struggle to run numerous simulations since the quality of the surrogate model relies on the amount of training data [28]. It is not surprising that such a modelling process still offers extensive overall savings in terms of the total number of computational simulations required. Therefore, surrogate modelling methods have been applied to probabilistic optimisation due to their advantages regarding computational efficiency. One drawback of surrogate modelling is that it sometimes requires a large number of simulations for training when the optimisation problem has many design variables. For example, if the optimisation problem involves complex and large-scale composite structures, the runtime for a single FEM model might extend beyond a few hours despite using high-performance computing resources. There are many types of surrogate modelling methods; some of the most commonly employed methods are Gaussian processes and artificial neural networks (ANNs).

An inclusive summary and theoretical knowledge of surrogate modelling, which is more specific to the work introduced in this monograph, can be found in Chapter 3.

1.4 Multi-Fidelity Modelling

Multi-fidelity modelling, which arises from surrogate modelling, has drawn significant attention among the optimisation research community since its introduction to computer science, particularly in the past three decades [29]. Multi-fidelity modelling aims to attain a high computational efficiency compared to the traditional surrogate modelling method. A general surrogate model is created using a high-fidelity model (HFM) that generates accurate solutions while being computationally expensive. In contrast, a multi-fidelity model is trained using both an HFM and a low-fidelity model (LFM), which is not accurate but computationally inexpensive. Multi-fidelity models, constructed by blending different fidelity models, can yield output solutions as accurate as those provided by general surrogate models using only HFMs while reducing computational costs to a similar level to that of LFMs. The ideal concept of multi-fidelity modelling involves using a small number of HFMs and a large number of LFMs to achieve a specific reduction in total training time. The fidelities defining the HFM and LFM depend on how the former can be simplified by the latter, such as in terms of physics and numerical accuracy. Multi-fidelity models are usually constructed using the surrogate modelling methods to achieve greater computational gains compared to those using high- and low-fidelity FEM models [30, 31].

One remarkable advantage of multi-fidelity modelling is its ability to deal with the computational challenge of probabilistic design optimisation for complex and large-scale composite structures, which traditional surrogate modelling methods cannot manage. Suppose a multi-fidelity model can provide acceptable accuracy with computational efficiency for the design of composite structures. In that case, engineers will be able to examine the statistical estimates of design uncertainties at the early design stages without resorting to expensive experiments or time-consuming simulations.

Prior to the work introduced in this monograph, research on multi-fidelity probabilistic design optimisation for composite structures had not been conducted in the engineering and science community. Given the importance of composite structures under the nonlinear post-buckling regime and thermomechanical loading, the critical need for probabilistic design optimisation to consider design uncertainties, and also the incredible benefits of surrogate and multi-fidelity modelling, a novel methodology that embraces these different approaches would provide a suitable answer to the technical challenges that the industry seeks to tackle.

A comprehensive explanation and theoretical background of multi-fidelity modelling can be found in a book by Forrester *et al.* [11]. An overview of the various multi-fidelity models, highlighting the work presented in this monograph, can be found in Chapter 3.

1.5 Overview of the Monograph

The objectives of this monograph are to describe the development of novel multi-fidelity modelling-based probabilistic design optimisation methods for composite structures, demonstrate these methods using the design problems of a mono-stiffened stringer composite panel, and broaden their application area, particularly for designing large-scale composite structures. The overall outline can be summarised in the following five sub-objectives:

(1) A multi-fidelity RBDO framework for composite structures integrated with surrogate modelling is developed. Multi-fidelity models that consist of both an HFM and an LFM defined by different FEM mesh sizes are constructed using response correction functions. Then multi-fidelity surrogate models are created using ANNs. The multi-fidelity RBDO framework involves the multi-fidelity modelling process and different reliability analysis methods, such as MCS, FORM, and SORM. This framework is demonstrated for the first time using the engineering examples of a mono-stringer stiffened composite panel, considering uncertainties in its geometry and the applied load. These examples evaluate the developed multi-fidelity RBDO framework regarding improvements in solution accuracy and computational time savings compared to traditional high-fidelity surrogate modelling techniques. The framework is presented in Chapter 4.

(2) A multi-fidelity modelling formulation covering different design spaces between an HFM and an LFM is developed. The main drawback of traditional multi-fidelity modelling methods, in which the HFM shares the same design space as the LFM, can be addressed by configuring the two to have a different number of design variables. The HFM only has a few design variables to reduce the number of high-fidelity FEM simulations for training. At the same time, the LFM explores the entire design space, sharing the design variables with the HFM. The multi-fidelity formulation incorporates multi-level optimisation into the modelling process to complement the limited information due to the HFM not carrying all design variables. This delivers more computational efficiency, followed by surrogate modelling based on ANNs, making it particularly suitable for large-scale problems with many design variables. This new multi-fidelity formulation is presented in Chapter 5.

(3) A multi-fidelity RDO framework for composite structures under the nonlinear post-buckling regime, where the HFM has a smaller number of design variables than the LFM, is developed. The developed multi-fidelity formulation is integrated with the RDO process. Then, it is demonstrated through two optimisation problems, deterministic

optimisation and RDO, of a mono-stiffened stringer composite panel undergoing mechanical shortening beyond its linear buckling limit. The design uncertainties in geometric parameters are considered, and MCS is used to predict the statistical characteristics through the Sobol sampling technique for each design point. The developed framework can improve accuracy and save computational cost over both conventional surrogate modelling and various multi-fidelity modelling methods. The comparison between robust and deterministic designs is highlighted by examining the variability of output responses, relying on the consideration of design uncertainties. The developed multi-fidelity RDO framework is presented in Chapter 5.

(4) A multi-fidelity modelling formulation utilising different sampling levels between an HFM and an LFM while considering nonlinear correlations between them is developed. This multi-fidelity formulation can offer acceptable multi-fidelity models that take care of the nonlinear correlations between different fidelity models in a complex structural problem. This new formulation involves both a nonlinear information fusion algorithm and multi-level optimisation. Specifically, the formulation enables the HFM to supervise a part of the entire design space using high-fidelity information collected densely from the selected design spaces. Simultaneously, the HFM provides high-fidelity information of other design variables collected sparsely without increasing the sampling size in the high-fidelity training dataset. The LFM explores the whole design space, enabling engineers to explore the solution spaces of other design variables associated with the selected design variables in the HFM. This novel multi-fidelity formulation is presented in Chapter 6.

(5) A multi-fidelity probabilistic optimisation framework for composite structures subjected to thermo-mechanical loading is developed. The developed multi-fidelity formulation is incorporated into the RBDO process. Then, it is demonstrated through a numerical example of a mono-stiffened stringer composite panel under mechanical and thermal loading for the first time. The RBDO process aims to maximise the critical temperature changes occurring due to inevitable mechanical shortening while ensuring the target reliability is met. In this example, the constructed multi-fidelity model carries 10 input parameters to consider the design uncertainties in both material properties and geometry during the optimisation process. The optimal designs are found

using successive high-fidelity corrections at the end of each optimisation level. The accuracy improvements and computational time savings are highlighted by a comparison with different traditional methods and a surrogate model computationally equivalent to the multi-fidelity model. The developed multi-fidelity optimisation framework is presented in Chapter 6.

Chapter 2

Fundamentals of Structural Optimisation

This chapter summarises the fundamental concepts on which the research presented in this monograph is based. The chapter introduces the basic optimisation methods, such as gradient methods, direct methods, and evolutionary methods, which can be used for probabilistic design optimisation. It then presents multi-objective optimisation for probabilistic design, including reliability-based design optimisation (RBDO) and robust design optimisation (RDO); it highlights a comparison between RBDO and RDO as well. The theory behind the consideration of design uncertainty is then described by showing how to define the uncertainty in structural design optimisation. Reliability analysis and robustness analysis are discussed based on different approaches to assessing uncertainty, such as Monte Carlo simulation (MCS), the first-order reliability method (FORM), and the second-order reliability method (SORM).

2.1 Optimisation Methods for Probabilistic Design

A vast number of optimisation methods have been developed in the field of structural optimisation over the past few decades. These optimisation methods aim to minimise specified objective functions by adjusting a set of design variables without violating any given design constraints. In general, these are not particular types of optimisation methods developed to achieve probabilistic design. However, some researchers have proposed efficient optimisation methods for probabilistic design that provide adequate performance accuracy and efficiency [32]. These methods can be categorised into several types, such as gradient-based methods, direct methods, and evolutionary methods. Gradient-based methods require the derivative information of problem functions to determine optimised search directions. Direct methods are derivative-free methods; they do not calculate the derivatives of functions in their directional search scheme. Evolutionary methods use the information of function values in the optimisation process. More details of these methods can be found in many works of literature on the topic of optimisation methods for probabilistic design [33–40].

2.1.1 *Gradient-based methods*

Gradient-based methods use the derivatives of problem functions to determine the search direction towards the optimum points. Many methods have been developed for identifying this optimal search direction, including simple steepest descent and conjugate gradient [41]. These methods decide on a direction using either the first- or second-order Taylor approximation, depending on the characteristics of the problem to calculate the derivatives at each step. The use of derivatives for the search direction scheme requires that the problem functions be twice continuously differentiable in the feasible design space to obtain more accurate values. At the same time, the design variables in the problem should be continuous within the design ranges to offer appropriate values. The main advantage of these methods is that they converge extremely quickly compared to other methods because they usually need only a small number of simulations to be iterated. However, gradient-based methods employ local information, such as function values and their gradients, during the search direction scheme. This could result in convergence to a local minimum that satisfies the given conditions. There have been many gradient-based methods developed so far, such as mixed integer sequential quadratic programming (MISQP), nonlinear programming, sequential quadratic programming (SQP), and large-scale generalised reduced gradient (LSGRG) [42]. Each method has its own characteristics depending on the problem, design space, CPU resources, features, etc. LSGRG is known for its potential for handling nonlinear design spaces and problems with many design variables, making it suitable for probabilistic design optimisation of large-scale problems. This method adapts the search direction to maintain the precision of active constraints while also following the constraints to improve the design. The method divides the gradient calculations of possible search directions to ensure parallelisation, in contrast to typical gradient-based methods.

2.1.2 *Direct methods*

As direct methods do not take the derivatives of the problem functions, they search for only certain points in a local design space to determine the direction and then decrease the size of the local searched area to achieve convergence. The typical direct methods include the Hooke–Jeeves method and the downhill simplex method [43]. The Hooke–Jeeves method investigates points around the current point through the perturbation of each design variable, one at a time, until an improvement is achieved. When there is no further improvement in design tracking the favourable direction, variable perturbation is gradually reduced until convergence criteria are satisfied. In contrast, the downhill simplex method is a geometrically intuitive algorithm that provides the capability to search in every direction by building a multi-dimensional body. As this algorithm proceeds, the simplex leads to a downward path towards the minimum region through a series of steps. This method has a reasonably high probability of discovering the global minimum when the initial steps are large. The initial simplex will then cover a broader range of the design space and reduce the possibility of getting stuck at a local minimum. Even though direct methods are intuitive and straightforward enough to implement, they show weak performance in the discontinuous design space. In addition, convergence can be extremely slow in complex, large-scale problems, and they may not even converge to a stationary point.

2.1.3 *Evolutionary methods*

Evolutionary methods are inspired by nature. The main characteristic of these methods is that they do not require the derivatives of problem functions but instead compute function values. This enables the problem functions to be non-differentiable and discontinuous across the entire design space, unlike the general requirement of gradient-based methods. The approach of evolutionary methods involves randomness in the search process, whereas gradient-based methods determine a specific search direction calculated using derivatives [44]. These methods seek to run a number of simulations without calculating derivatives to explore the entire design space as much as possible and develop solutions gradually in comparison with previous ones. Many evolutionary methods using this idea have been developed and demonstrated so far, including genetic algorithms (GAs), particle swarm optimisation (PSO), and non-dominated sorting genetic algorithm-II (NSGA-II) [40]. These methods commonly provide much better coverage of the design space. Thus, they ensure a global minimum point concerning the objective functions while not getting trapped in different local minimum points. Given that these gradient-free optimisation methods do not rely on derivative tests to reach a minimum but rather explore the design space to discover the best fitness value, the optimal point could be evaluated using mathematical conditions such as the Karush–Kuhn–Tucker (KKT) conditions. An additional helpful feature of the methods is that they can manage various types of problems, including those with continuous, integer, and mixed variables, since they are not concerned with differentiability conditions. A major drawback is that they are more computationally expensive than gradient-based methods since a large number of simulations are required to discover the area containing the global minimum point. However, this drawback can be resolved through parallelisation, one of the benefits offered by these methods.

It is not surprising that the GA is the most well known among the evolutionary methods. The primary mechanism of this algorithm involves mimicking genetic operations that consist of certain critical parameters, such as population and generation [35]. The population is a set of design points and also represents possible solution points. The generation shows how many iterations the GA carries out during the optimisation process. The total number of design points considered depends on these two parameters. The GA allows the population of design points to be gradually improved over consecutive generations. The GA process begins with an initial population, which is randomly selected from across the whole design space. In this set of design points, a subset is chosen randomly, and then random processes generate new design points using the selected subset. The following sets of design points provide better fitness values as a result of utilising a subset of the previous set. This process is terminated until a stopping condition or a maximum allowable iteration is satisfied. This process is accomplished through three main genetic operators: selection, crossover, and mutation. The selection process aims to reproduce an old design point depending on fitness values for the generation of a new population. In the crossover process, the selected design points of the new population exchange characteristics among themselves. The mutation process permits additional randomness to safeguard the process by mutating a design variable at each design point using another random value. The NSGA-II is a more advanced GA that aims to deal with multi-objective problems [45]. It considers each objective separately, while the standard genetic operation of mutation and crossover are conducted.

By the end of the optimisation process a Pareto front is generated by selecting feasible non-dominated design points, where each design point obtains the best combination of objective values. An improvement in one objective is not possible without sacrifices on one or more of the other objectives. This algorithm has been considered a method for multi-objective probabilistic design optimisation due to its excellent performance.

2.2 Multi-Objective Probabilistic Optimisation

Multi-objective optimisation, which is also called multi-criteria optimisation or vector optimisation, is an optimisation process capable of systematically and simultaneously minimising several objective functions. In this process, the method of determining a solution concerning several objective functions should be defined, unlike single-objective optimisation, which seeks to find a solution by minimising a single objective function [45]. The different objective functions of this process may conflict with each other, which means an improvement in one objective might often leads to degradation in others. Several constraints can also be considered, which feasible solutions having every optimal solution must satisfy. Equation (2.1) shows a general form of multi-objective optimisation:

$$\min_{\mathbf{d}} \quad F(\mathbf{x}) = (f_1(\mathbf{d}, \mathbf{x}), \ldots, f_p(\mathbf{d}, \mathbf{x}))$$

$$\text{subject to} \quad g_i(\mathbf{d}, \mathbf{x}) \leq 0, \quad i = 1, \ldots, I$$

$$h_j(\mathbf{d}, \mathbf{x}) = 0, \quad j = 1, \ldots, J \tag{2.1}$$

$$\text{where} \quad F : \mathbb{R}^n \to \mathbb{R}^p$$

where F is a vector including all p objective functions, $\mathbf{d}$ and $\mathbf{x}$ are the vectors of the design variables and the whole optimisation variables, g_i and h_j are inequality and equality constraint functions, and $\mathbb{R}^n$ and $\mathbb{R}^p$ are the vectors of inputs and objectives, respectively.

The optimal solutions in multi-objective optimisation problems can be characterised by a mathematical concept of dominance or partial ordering [46]. The idea of dominance between two solutions, d_1 and d_2 in $\mathbb{R}^n$, is explained using equation (2.2). If d_1 dominates d_2, two conditions should be necessarily considered. First, d_1 is not worse in any of the objectives. The value of the objective function for d_1 is less than or equal to the value of the objective function for d_2, regarding all the required objective functions. Second, there must be at least one objective for which d_1 is rigorously greater:

$$\forall i \in \{1, \ldots, p\}, \quad f_1(d_1) \leq f_2(d_2)$$
$$\exists i \in \{1, \ldots, p\}, \quad f_1(d_1) < f_2(d_2)$$

(2.2)

This process gives rise to a set of optimal solutions that is defined as *Pareto optimality*, i.e., a feasible vector d^* is Pareto optimal if the vector d^* is not dominated by any other possible solution. Equation (2.3) represents that d^* is Pareto optimal if no objective can be improved without sacrificing more than one objective of the other objectives. The set of Pareto optimal solutions is prescribed as the *Pareto optimal set*, and it provides the optimal solutions of multi-objective optimisation problems:

$$\text{Pareto optimal set} = \{d^* \in F \mid \nexists d \in F : F(d) < F(d^*)\} \qquad (2.3)$$

The main difference in multi-objective probabilistic optimisation is that the optimisation process considers the uncertainties of design variables throughout the entire structural lifecycle. This consideration enables the final design to become a more reliable or robust design, depending on the defined objectives and constraints. There are two types of probabilistic optimisation: RBDO and RDO [16, 24]. RBDO carries the probability of failure as a constraint, whereas RDO considers the means and standard deviations of objective functions. These two types are introduced in the following sections in detail.

2.2.1 *Reliability-based design optimisation*

In general, RBDO integrates reliability analysis with deterministic optimisation so that the optimisation process assesses the design constraints resulting from the uncertainties in random design variables. This optimisation process ensures that the final design satisfies a specific probabilistic constraint up to a prescribed reliability level. When an optimisation problem focuses on the occurrence of a disastrous failure of a structural system, the optimisation problem is defined as RBDO [9, 15, 47]. In RBDO, reliability analysis evaluates a limit state function to calculate the probability of failure. A constraint is imposed to ensure that failure does not exceed an adequate critical value. This limit state function is related to the constraints used in deterministic optimisation, with the distinction that constraints may be violated at an acceptable level of probability. A typical

RBDO problem can be expressed as

$$\text{minimise} \quad F(d)$$

$$\text{subject to} \quad g_m(d) \leq 0, \quad m = 1, \ldots, M \tag{2.4}$$

$$P[G_n(d, X)] - P_{f,n} \leq 0, \quad n = 1, \ldots, N$$

where d and X are the vectors of the design variable and random variable, respectively, and F represents the objective functions. g_m is the mth deterministic constraint, and G_n represents the nth probabilistic constraint. $P[\cdot]$ implies the probability of the constraint being satisfied, and $P_{f,n}$ is the acceptable probability of failure. $P_{f,n}$ mainly indicates the prescribed reliability level β_t when a normal distribution represents the random variables.

The reliability analysis in this optimisation process is an essential part of determining the probability of failure, which can be predicted using MCS, FORM, or SORM [8]. MCS is based on different sampling methods, while FORM and SORM exploit the derivatives of the limit state function. This reliability assessment, considering design uncertainties, requires thousands of function evaluations, leading to significant computational costs, which, however, is not a critical issue for deterministic optimisation. Different advanced numerical techniques have been developed to address the computational cost that presents a challenge with current technologies.

2.2.2 *Robust design optimisation*

RDO combines the concept of robustness with deterministic optimisation, which may produce over-optimised design solutions caused by the ignorance of design uncertainties [48, 49]. A robust design aims to improve the quality of products by accounting for unexpected deviations caused by variations at different phases of a structure's lifecycle. This means that the structural performance of the robust design should be less sensitive to random variations. Even though the design solutions found through a conventional deterministic optimisation approach work well at design points, they could present inferior performance close to or away from the design points. It is not surprising that the design solution which results from deterministic optimisation will not be the robust solution providing minimum sensitivity to the design uncertainties. Structural performance is represented by objective functions or constraints, and it may exhibit wide scatter across different service phases. This scatter may considerably decrease structural quality and lead to deviations from the required performance. They may also increase the structural costs associated with inspections, repairs, and maintenance. In that sense, well-optimised design solutions provide structures that reduce operating costs as well as the scatter of structural performance. These structures function consistently in the presence of unexpected variations during the overall service lifecycle. The robustness of structures is one of the essential characteristics that must be considered in the design stages to reduce the scatter of structural performance.

One possible way is to reduce or exclude the spread of the input parameters, which may either be nearly impossible or increase the structure's total costs. Another way is to discover a design in which the structural performance is insensitive to the deviation of input parameters without removing the sources of the parameters' variations. Such a robust structural design approach describes the design quality of the structure using the mean value and variation of the structural performance. An evident approach is to specify the optimality requirements using the expected values of response performance. However, the design determined by the minimum expected value of the objective functions may still be susceptible to variations in the probabilistic input parameters, which exhibit uncertainties. This draws attention to the need for a robust structural design that balances mean performance against some measure of its variability. This design is achieved through multi-objective optimisation, which involves a trade-off between the two [4]. The general mathematical formulation of RDO can be given as

$$
\begin{aligned}
\min \quad & \{\mu(f(X)), \sigma(f(X))\} \\
\text{subject to} \quad & g_i(X) \leq 0, \quad i = 1, \ldots, I \\
& h_j(X) = 0, \quad j = 1, \ldots, J \\
& \sigma(f(X)) \leq \sigma_M^+ \\
& x_l^{(L)} \leq x \leq x_l^{(U)}, \quad l = 1, \ldots, L
\end{aligned}
\tag{2.5}
$$

where X is the vector of the design and random variables and $\mu(f(X))$ and $\sigma(f(X))$ are the first (mean) and second (standard deviation) statistical moments of the objective function, respectively. g_i is the ith inequality constraint, h_j is the jth equality constraint, σ_M^+ is the upper limit for the standard deviation of the structural performance, and $x_l^{(L)}$ and $x_l^{(U)}$ are the lower and upper bounds for the lth design variable, respectively.

2.2.3 *Reliability-based design vs. robust design*

As discussed in the previous section, RBDO and RDO aim to incorporate design uncertainties into the design optimisation process. These
two optimisation approaches presume that design hazards are defined by
combining the probability of an undesired event and its consequences.
They put forth different design domains, depending on the purpose of
each optimisation. Figure 2.1 highlights the design domains of RBDO
and RDO using two principal elements: the event's consequences and
its frequency. A design could target a structural system encountering an
extremely severe environment that leads to structural failure and collapse.
This design is placed in the domain of RBDO, where the concept of the
reliability level is considered by introducing a critical constraint. In contrast,
a design could be optimised such that a given circumstance does not
cause significant failure but deteriorates the structural performance due
to frequent fluctuations. In terms of design domain, RBDO is more likely
associated with structural safety under extreme design conditions, whereas
RDO relates to structural performance under operating conditions.

Both RBDO and RDO combine the prediction of possible influences
caused by design uncertainties with the optimal design process; however,
these two approaches have different interest areas in relation to the
distribution of the objective functions. RBDO aims to meet reliability
requirements based on the probability distributions of the random design

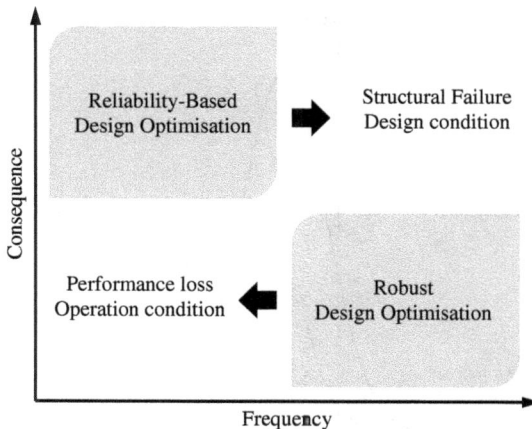

Fig. 2.1. Design domains depending on design target.
Source: Ref. [50].

variables [15, 32]. The objective functions of RBDO should be minimised without violating the probability constraints. RDO reduces variability in structural performance [4, 51]. This approach includes minimising the variance of objective functions. Simultaneously, constraints can also be characterised by the allowable standard deviations. In RBDO, stochastic analysis predicts the likelihood of extreme events, as illustrated by the tail of the objective function's distribution. The reliability analysis, which is the main part of RBDO, measures the reliability index associated with the probability of failure [50]. However, a high computational cost is required to compute the reliability of each design point through this process. Stochastic analysis in RDO calculates the effects of statistical moments at different design points. The robustness assessment of RDO concerns the variation of structural performance regarding unanticipated events. Figure 2.2 highlights the aim of RBDO, which is to reduce the probability of rare extreme events in the tail of the probability distribution function. The limit state function serves as the primary constraint, acting as the boundary between the safe and failure regions. RBDO generally diminishes the tail area across the limit state function so that the structure achieves the required reliability level. In contrast, Figure 2.3 illustrates how RDO ensures an insensitive design. This design process reduces the variance of the objective function, resulting in a narrower distribution around the mean value of the function. Both clearly show how the two approaches offer

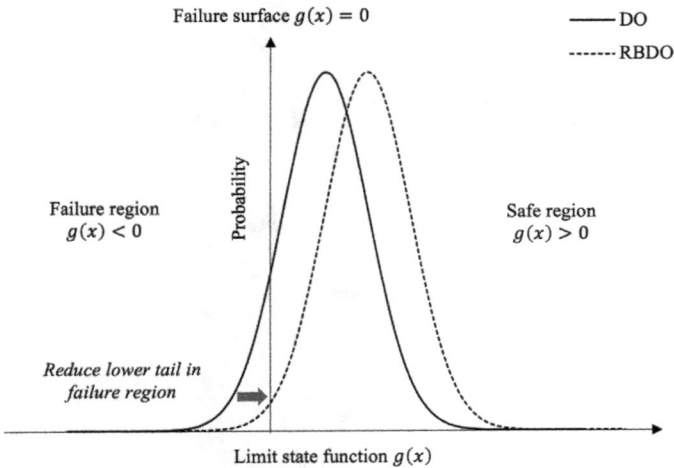

Fig. 2.2.   RBDO strategy.

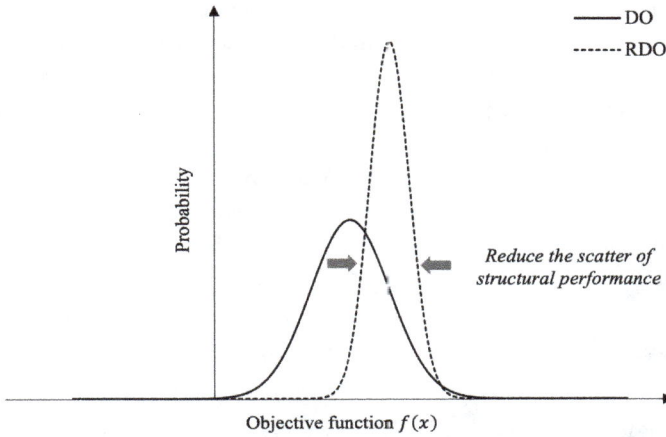

Fig. 2.3.   RDO strategy.

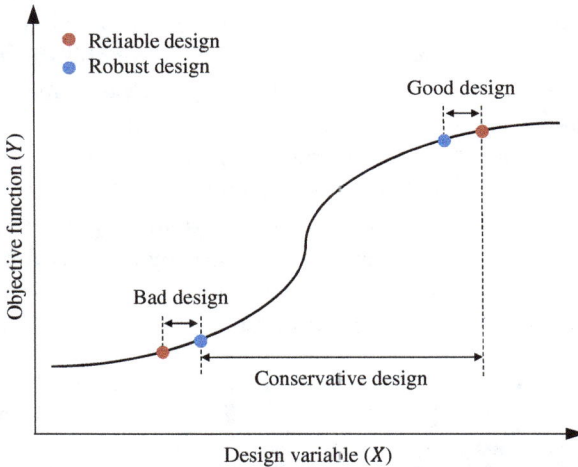

Fig. 2.4.   Reliable design and robust design.

different design solutions compared with those of traditional deterministic optimisation. When robust and reliable designs are found, it is meaningful to evaluate whether the final chosen design is feasible from an engineering perspective.

Figure 2.4 highlights how a feasible design can be obtained using the probabilistic design approach. If the objective function of a robust design (under service conditions) is much smaller than that of a reliable design (under extreme conditions), the final design will resemble a conservative design resulting from assuming large safety factors. If the objective function of a robust design is adjacent to that of a reliable design, the final design will be a good design that reduces the difference in structural performance between the service and extreme conditions. Suppose the objective function of a robust design is better than that of a reliable design. This does not imply that the mean value of the objective function for the operating conditions is higher than that for the extreme conditions.

2.3 Design Uncertainty Considerations

As design requirements become increasingly complex and comprehensive, many design approaches have been widely developed to meet the need for advanced probabilistic assessment techniques considering the design uncertainties. In particular, the consideration of these uncertainties enables the design approaches to deal with the random nature of design parameters. Moreover, it ensures a superior design quality compared to other design approaches that are based on safety factors. Probabilistic designs that account for design uncertainties safeguard against performance deviations caused by system variations. Such designs can resolve shortfalls in knowledge about the spread of structural performance, which is the primary concern in a design based on safety factors [5, 52]. As can be seen in Figure 2.5, the sources of uncertainty are commonly categorised by the product's lifecycle, such as design, manufacturing, service/operation, and ageing. In the design stage, the uncertainties are usually inaccuracies in models and insufficient information stemming from different levels of

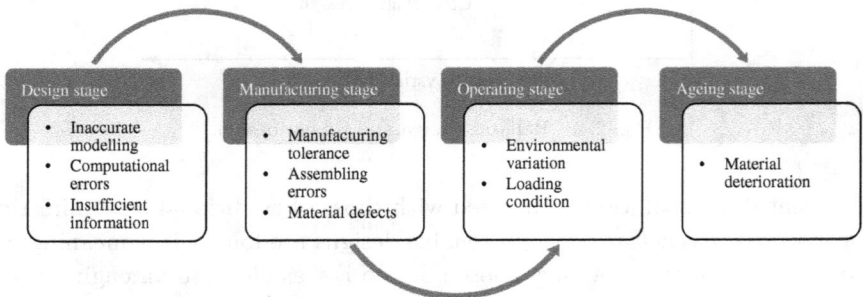

Fig. 2.5.　Uncertainties through the entire lifecycle.

understanding the structural system. These uncertainties in models and information are associated with the precision of model creation and the extent to which engineers are able to understand the system, respectively. Tolerances and material defects are considered uncertainties in the manufacturing stage, which follows the design stage. The tolerances include changes in geometry during the manufacture of the final design or assembly in the production lines. In particular, material defects are crucial for manufacturing composite structures since misalignments in fibre orientation and the presence of voids in the matrix influence the overall mechanical properties of composite structures. The operation stage has different uncertainties, such as environmental variations and loading conditions, which are more likely to be associated with structural performance during a product's service life. Finally, the decay of material properties may lead to the efficiency loss of the structural system during the ageing stage.

Conventional design approaches to optimisation, which are collectively known as deterministic design, do not explain any uncertainties in the structural system. These design approaches may provide an over-optimised design, which may not present adequate capability close to or outside of the design points, despite presenting perfect performance at the design points. In contrast, modern probabilistic design approaches to optimisation account for the uncertainties that can be expected throughout the entire lifecycle. This consideration ensures that the design approaches can predict how much the uncertainties affect the design objectives and offers more statistically reliable and robust designs. There are two analysis types to assess the uncertainties associated with the whole lifecycle, namely reliability and robustness [53]. In RBDO, reliability analysis is an essential process to predict the probability of failure caused by design uncertainties. In general, a structural design produced by a deterministic design approach based on safety factors has a greater probability of failure compared to a probabilistic design that takes the uncertainties into account. This ensures that the reliable design provides a certain confidence level without violating a prescribed probabilistic constraint across the whole lifecycle since RBDO examines the effects of uncertainties. Conversely, the RDO process includes a robustness analysis to calculate the statistical variance in the structural performance. This analysis enables the optimisation process to minimise the sensitivity of the objective function concerning arbitrary changes in the random variable in the system. These two analyses are discussed further in the following sections.

2.3.1 *Reliability analysis*

Reliability analysis is one of the approaches to consider uncertainties across the entire lifecycle of products [54]. It becomes more significant in the field of structural design because it provides many benefits. This analysis mainly allows engineers to realise how the uncertainties in different design variables impact the probability of failure of the structures they designed. At the same time, it enables designers to understand where the most critical design region is in the whole design space and improve the reliability of the system. There are two types of methods, depending on how they predict a structure's reliability, namely statistical methods and non-statistical methods [55]. MCS represents a statistical method, while FORM and SORM fall under the category of non-statistical methods. FORM and SORM utilise the first-order and second-order Taylor series expansions, respectively, to approximate a limit state function.

2.3.1.1 *Theory*

Reliability analysis considers the uncertainties of design variables and computes the probability of structural failure corresponding to the reliability index. The reliability analysis evaluates whether a limit state function, which is a prescribed constraint, exceeds a specific value. The limit state function represents a state in which a structure can no longer perform its design purpose since it experiences conditions exceeding particular allowable limits. If the probability of failure regarding the limit state function is greater than the specified required value, the structure does not provide acceptable confidence associated with the reliability level. The limit state can be classified into two types: ultimate and serviceability limit states [8]. The ultimate limit state is represented by the phenomenon of structural collapse, such as corrosion, fatigue, deterioration, the plastic mechanism, progressive collapse, and fracture. These limit states have a very low possibility of occurrence, but their consequences can be dangerous. In comparison, the serviceability limit state is related to the disruption of structures. For example, this may include excessive deflection, excessive vibration, drainage, leakage, and local damage. Such limit states have a higher possibility of occurrence, albeit with less harmful consequences. In general, the limit state function shows the margin of safety between the resistance and load of structures. The limit state function can be defined as

$$g(Z) = R(X) - S(X) \tag{2.6}$$

$$P_f = P[g(Z) < 0] \tag{2.7}$$

where $g(Z)$ is the limit state function, P_f is the probability of failure, Z is a vector of design variables that influences the limit state function, while X ($X \subseteq Z$) is a vector of design variables that affects R and S, with R being the resistance and S being the loading of the structure.

If the value of $g(Z)$ is less than zero, the structure is in the failure region. If the value of $g(Z)$ equals zero or is larger than zero, the structure is in the failure surface or the safe region. respectively. The reliability index β, which indicates a confidence level without violating the prescribed limit state, is expressed as

$$\beta = \frac{\mu_g}{\sigma_g} \tag{2.8}$$

where μ_g and σ_g are the mean and standard deviation of the limit state function, respectively.

The reliability index is the distance of the mean value of the limit state function from the safe surface, and it is an appropriate measurement of reliability. When the limit state function is normally distributed, the probability of failure is described as

$$P_f = P\{g(Z) < 0\} = \int_{-\infty}^{0} f_Z(z)\,dz = 1 - \Phi(\beta) = \Phi(-\beta) \tag{2.9}$$

where $f_Z(Z)$ is the joint probability density function of Z, P_f are calculated by integrating over the failure region $g(Z) < 0$, β is the reliability index, and $\Phi(\cdot)$ is the standard normal cumulative distribution function.

All design variables in equation (2.9) are presumed to be independent of one another. It can be challenging to calculate the integral in the equation if the failure surface is nonlinear or if there are many random variables in the limit state function. When the reliability index is calculated, the probabilities of failure and success are obtained. There are three numerical methods, namely MCS, FORM, and SORM, that can be employed to carry out reliability analysis. MCS is the most straightforward among the three methods. It evaluates the limit state function using randomly sampled design points from the distribution of the random design variables in the limit state function. In contrast, FORM and SORM simplify the limit state function using the first-order and second-order Taylor series expansions, respectively, thereby approximating the failure surface. Thus, FORM and SORM require much shorter calculation times, but they are not as accurate as MCS.

2.3.1.2 *Monte Carlo simulation for reliability analysis*

MCS is a simple random sampling method or a statistical trial method based on randomly generated sampling sets for design variables [9]. It is a powerful mathematical method for specific events resulting from stochastic processes. MCS consists of the creation of random design variables, followed by the statistical analysis of their outcomes. When a distribution type for random design variables is determined, MCS builds up a sampling set from the determined distribution. Then, MCS runs simulations using the created sampling set. There are several parameters for a random sample, such as the number of sampling points and distribution type. The basic process of MCS is extended to the reliability analysis of structures. First, sampling sets of random design variables are collected using the probability density function. Then, the mathematical model of the limit state is set up, and this model evaluates failures in the sampled sets. Next, many simulations are conducted using the created sampled sets of the random design variables. Finally, the probabilistic characteristics of the structural response are estimated. As can be seen in equation (2.6), the limit state function $g(Z)$ to be evaluated is made up of $R(X)$ and $S(X)$, which are usually finite element method (FEM) or boundary element method (BEM) in the area of structural design bearing unknown probability distributions. If a vector X consists of random design variables with known probability distributions that affect S and R, the variables in X are sampled by their respective probability distributions. Then, the outcomes of $S(X)$ and $R(X)$ concerning the sampled X are obtained. This process is continued repeatedly until the probability distribution of $g(Z)$ is estimated precisely. If a total sampling number, N_{Total}, for the random design variables in X has been accomplished, the number of samples for which $g(Z_i) < 0$ or $S(X_i) > R(X_i)$ (where $i = 1, 2, \ldots, N_{\text{Total}}$) can be found, where X_i and Z_i are the ith sample for the random design variables in X and Z, respectively. This number is called N_{Failure}. When N_{Total} simulations are carried out, the probability of failure is computed by

$$P_f = \frac{N_{\text{Failure}}}{N_{\text{Total}}} \tag{2.10}$$

The probability of success, P_s, also known as reliability, can be calculated as

$$P_s = 1 - P_f \tag{2.11}$$

MCS is generally used to validate whether different approximation methods, such as the FORM and SORM, offer acceptable results. This process provides the most accurate estimation through a large number of simulations.

2.3.1.3 *First-order reliability method*

There are several ways to approximate the limit state function using the Taylor expansion. The FORM approach begins with the first-order second moment (FOSM) method; it has subsequently been developed into the Hasofer–Lind (HL) method and the Hasofer–Lind and Rackwitz–Fiessler (HL-RF) method [8]. HL-RF method, generally known as FORM, is used in this work.

The FOSM method, also called the mean value FOSM (MVFOSM) method, simplifies the functional relations and mitigates the complexities in calculating the probability of structure failure. The meaning of 'first-order' is the first-order expansion of the limit state function. Random design variables for reliability analysis are defined by the first moment (mean) and second moment (variance). As shown in Figure 2.6, the MVFOSM method approximates the limit state function using the first-order Taylor series expansion at the mean value. The approximation at the mean value using

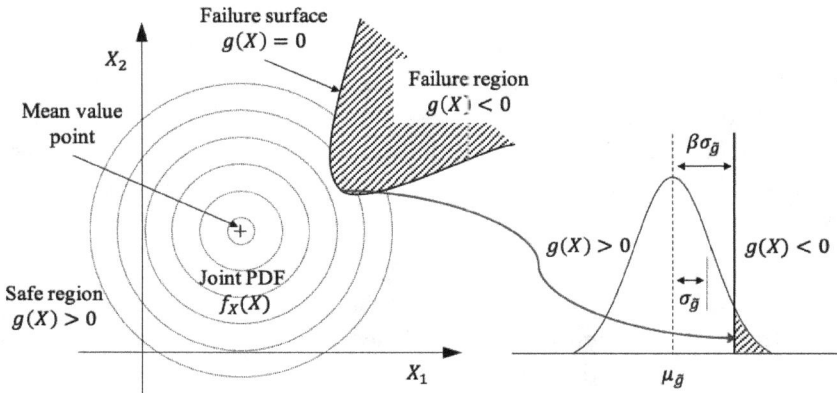

Fig. 2.6.   Mean value first-order second moment method.

statistically independent variables X is expressed as

$$\tilde{g}(X) = g(\mu_X) + \nabla g(\mu_X)^T (X_i - \mu_X) \tag{2.12}$$

where $\mu_X = \{\mu_{x_1}, \mu_{x_2}, \ldots, \mu_{x_n}\}^T$, and $\nabla g(\mu_X) = \{\frac{\partial g}{\partial x_1}(\mu_X), \frac{\partial g}{\partial x_2}(\mu_X), \ldots, \frac{\partial g}{\partial x_n}(\mu_X)\}^T$ is the gradient of the limit state function.

The mean value of the approximate limit state function $\tilde{g}(X)$ is written as

$$\mu_{\tilde{g}} = \mathbb{E}[g(\mu_X)] = g(\mu_X) \tag{2.13}$$

The standard deviation of the approximate limit state function is given as

$$\sigma_{\tilde{g}} = \sqrt{\mathrm{Var}[\tilde{g}(X)]} = \sqrt{(\nabla g(\mu_X)^T \, \mathrm{Var}(X))^2} = \sqrt{\left[\sum_{i=1}^{n} \left(\frac{\partial g}{\partial x_i}(\mu_X) \right)^2 \sigma_{x_i}^2 \right]} \tag{2.14}$$

Finally, the reliability index β is calculated as

$$\beta_{\mathrm{MVFOSM}} = \frac{\mu_{\tilde{g}}}{\sigma_{\tilde{g}}} \tag{2.15}$$

As can be seen from equation (2.15), it is the same as equation (2.8) if the limit state function is linear. If the limit state function is not linear, the approximate limit state surface is computed by linearising the original limit state function at the mean value.

The reliability index corresponding to the probability of failure is a mathematical optimisation indicator used to search for a point on the failure surface, $g(X) = 0$, representing the shortest distance from the origin to the surface in the standard normal distribution. This concept was enhanced by Hasofer and Lind through the Hasofer and Lind transformation. Here, a vector X of random design variables is transformed from basic variables into a set of normalised and independent variables U. The standardised form for this transformation is defined as

$$u_i = \frac{x_i - \mu_{x_i}}{\sigma_{x_i}} \tag{2.16}$$

where μ_{x_i} and σ_{x_i} are the mean value and the standard deviation of x_i, respectively. The mean and standard deviation of u_i are zero and one (standard normal distribution), respectively.

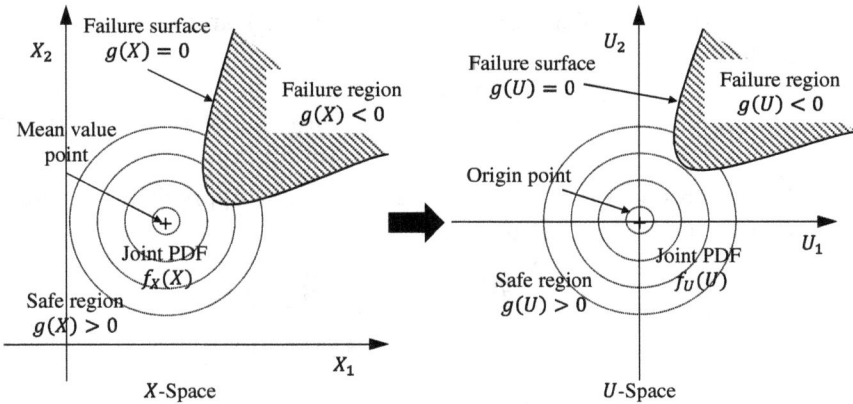

Fig. 2.7.   Failure surface transformation from X-space to U-space.

Through this formulation, the mean value of the original space (X-space) is moved to the origin of the standard normal space (U-space). The failure surface, $g(X) = 0$, in X-space is transformed into the associated failure surface, $g(U) = 0$, in U-space, as illustrated in Figure 2.7. The reliability index, β, is the shortest distance from the origin to the failure surface, $g(U) = 0$:

$$\beta = \min_{U:g(U)=0} \sqrt{U^T U} \tag{2.17}$$

The value of the reliability index is the same, not only for the true failure surface but also for the tangent hyperplane at the design point. The improvement in the HL iteration method changes the expansion location from the mean value to the most probable failure point (MPP). Let's assume the limit state function to be n-dimensional normally distributed with random design variables X:

$$g(X) = g(\{x_1, \ldots, x_n\}^T) = 0 \tag{2.18}$$

Based on the formulation of equation (2.16), the limit state function in X-space is mapped into U-space:

$$g(U) = g(\{\sigma_{x_1} u_1 + \mu_{x_1}, \ldots, \sigma_{x_n} u_n + \mu_{x_n}\}^T) = 0 \tag{2.19}$$

The first-order Taylor series expansion of $g(U)$ at the MPP, U^*, is expressed as

$$\tilde{g}(U) = g(U^*) + \sum_{i=1}^{n} \frac{\partial g}{\partial u_i}(U^*)(u_i - u_i^*) \tag{2.20}$$

The shortest distance from the origin to the approximated failure surface in Figure 2.7 is defined as

$$OP^* = \beta = \frac{g(U^*) - \sum_{i=1}^{n} \frac{\partial g(U^*)}{\partial x_i}\sigma_{x_i} u_i^*}{\sqrt{\sum_{i=1}^{n}\left(\frac{\partial g(U^*)}{\partial x_i}\sigma_{x_i}\right)^2}} \tag{2.21}$$

The direction cosine, which is called the sensitivity factor, is given as

$$\cos\theta_{u_i} = \cos\theta_{x_i} = -\frac{\frac{\partial g(U^*)}{\partial U_i}}{|\nabla g(U^*)|} = \frac{\frac{\partial g(x^*)}{\partial x_i}\sigma_{x_i}}{\sqrt{\sum_{i=1}^{n}\left(\frac{\partial g(U^*)}{\partial x_i}\sigma_{x_i}\right)^2}} = \alpha_i \tag{2.22}$$

The coordinates of the next point in U-space are calculated as

$$u_i = \frac{x_i - \mu_{x_i}}{\sigma_{x_i}} = OP^* \cos\theta_{x_i} = \beta\cos\theta_{x_i} \tag{2.23}$$

The corresponding coordinate in X-space is determined by

$$x_i^* = \mu_{x_i} + \beta\cos\theta_{x_i} \tag{2.24}$$

The procedure of the HL iteration method is shown in Figure 2.8.

In the HL method, random design variables should be represented as normally distributed. If the random design variables are not normally distributed, an additional transformation should be considered to obtain their normal distribution. Rackwitz and Fiessler proposed the transformation, and this advanced method is the HL-RF method. Once the distribution of design variables is transformed into a normal distribution, the HF-RF method becomes identical to the HF method, as shown in Figure 2.9.

2.3.1.4 *Second-order reliability method*

In general, FORM provides acceptable results when the limit state surface has only one shortest point and is linear near the design point. If the failure surface shows high nonlinearity, the reliability index and the probability of failure estimated using the FORM might not yield acceptable and accurate results. To deal with this challenge, SORM, as shown in

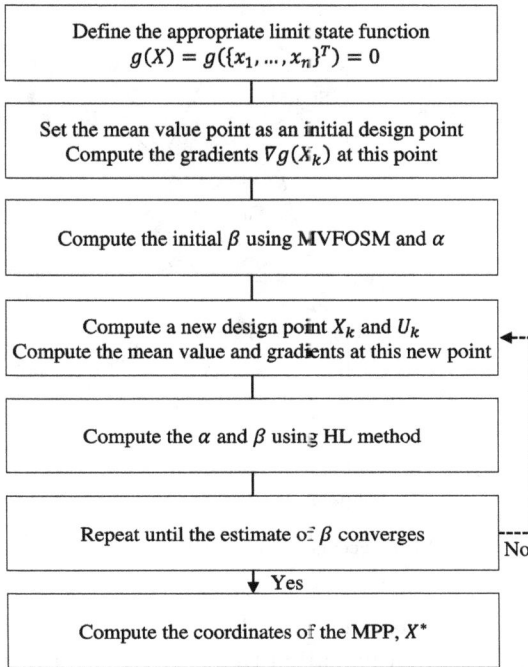

Fig. 2.8.   Procedure of the Hasofer–Lind iteration method.

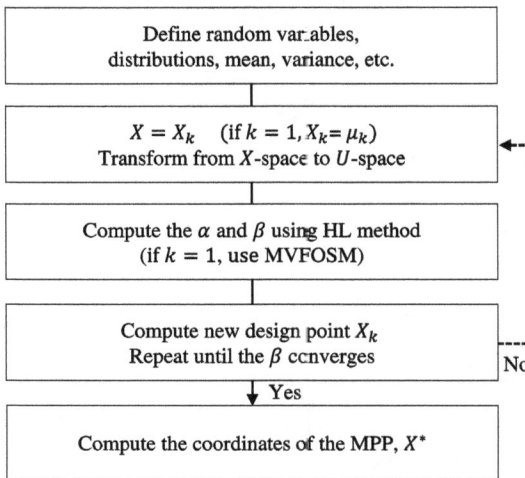

Fig. 2.9.   Procedure of the Hasofer–Lind and Rackwitz–Fiessler method.

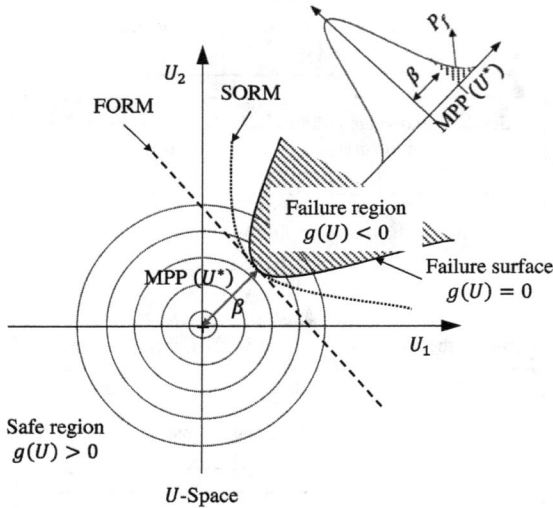

Fig. 2.10.   Second-order reliability method.

Figures 2.10 and 2.11, uses the second-order Taylor series expansion to obtain a more precise approximation that can replace the failure surfaces of the original function.

The second-order approximation of the limit state function, $g(U) = 0$, is derived using the second-order Taylor series expansion at the MPP:

$$\tilde{g}(U) = g(U^*) + \nabla g(U^*)^T(U - U^*) + \frac{1}{2}(U - U^*)^T \nabla^2 g(U^*)(U - U^*)$$

$$(2.25)$$

where $\nabla^2 g(U^*)$ is the symmetric matrix of the second derivative of the limit state function.

When equation (2.25) is divided by $|\nabla g(U^*)|$ and $g(U^*)$ is zero, the equation can be expressed by

$$\tilde{g}(U) = \alpha^T(U - U^*) + \frac{1}{2}(U - U^*)^T B(U - U^*) \qquad (2.26)$$

where $\alpha = \frac{\nabla g(U^*)}{|\nabla g(U^*)|}$ and $B = \frac{\nabla^2 g(U^*)}{|\nabla g(U^*)|}$

The matrix B is called the Hessian matrix, which is a square matrix of second-order partial derivatives of the limit state function. This Hessian matrix leads to substantial computational efforts for the reliability analysis. An orthogonal matrix H is introduced in the form of $Y = HU$ to carry

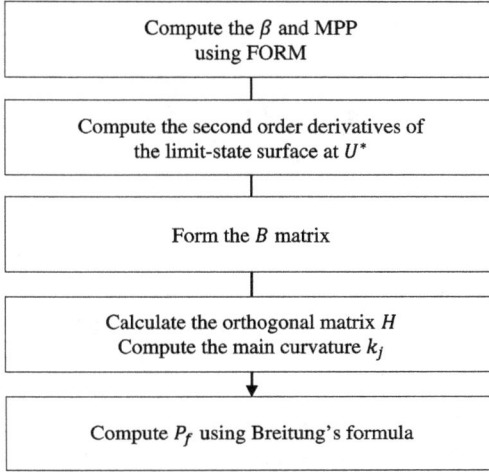

Fig. 2.11.   Procedure of the second-order reliability method.

out the rotation from the standard normal U-space to the rotated new standard normal Y-space. The details of the orthogonal matrix are found in textbooks on reliability analysis [8]. Equation (2.26) in U-space can be transformed into Y-space using the definition of direction cosine and the orthogonal matrix H, as follows:

$$\tilde{g}(Y) = -y_n + \beta + \frac{1}{2}(H^{-1}Y - H^{-1}Y^*)B(H^{-1}Y - H^{-1}Y^*) \qquad (2.27)$$

The final matrix in equation (2.27) can be given using the orthogonal matrix, H:

$$\tilde{g}(Y) = -y_n + \beta + \frac{1}{2}(Y - Y^*)HBH^T(Y - Y^*) \qquad (2.28)$$

$$y_n = \beta + \frac{1}{2}\sum_{i=1}^{n-1} k_i \left(y_i'\right)^2 \qquad (2.29)$$

where k_i indicates the curvature of the response surface at the MPP, and the major computational effort is caused by calculating the second derivatives of the limit state function at the MPP.

In particular, if the finite difference method (FDM) is considered for calculating the gradient of the limit state function, the massive computational time required might influence the efficiency of reliability analysis. There are two methods to compute the probability of failure using the SORM: the Breitung formulation, as expressed in equation (2.30) and Tvedt's formulation [8, 9]:

$$P_f = \Phi(-\beta) \prod_{j=1}^{n-1} (1 + k_j\beta)^{-1/2} \tag{2.30}$$

2.3.2 *Robustness analysis*

The concept of robustness is utilised to quantify the quality of a product in manufacturing engineering. When this concept is incorporated into the deterministic optimisation process, it ensures that the optimal solution is a robust design. The variation during operating or service conditions causes unexpected performance deviations from the initial design. These deviations may cause a decrease in the quality of structures as well as an increase in various costs associated with monitoring, inspection, repairs, and maintenance. Robustness analysis has gained attention in structural design to address the issues of quality loss and the increase in operational costs. The most straightforward way is to either decrease or eliminate the scatter of the input parameters; however, that is not feasible. The other way is to produce structures that are less sensitive to the variations in the input parameters.

2.3.2.1 *Theory*

The fundamental of robustness in structural design is that the optimal design is determined not only by the minimum mean value of the objective function but also by how much the structural response is scattered about the mean value [3]. The global optimal design in deterministic optimisation may be more sensitive than a local design, although the optimisation process finds the minimum mean value of the objective function. Figure 2.12 highlights the concept of robustness using the distribution of the objective function. The figure shows two different distributions concerning the same objective function that must be minimised through the optimisation process. Each curve illustrates the frequency of the objective function, considering the random perturbation to its mean value. In the figure, two curves representing different designs have their own mean and variance values. Even though 'Design A' presents a smaller mean value of the objective function, its dispersion level about the mean value is larger than that of 'Design B'. In comparison, the mean value of 'Design B' is greater than that of 'Design A', while 'Design B' offers a narrower distribution

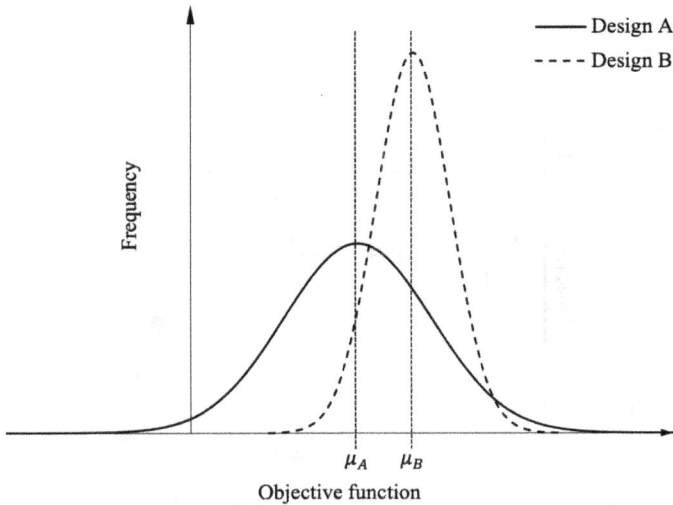

Fig. 2.12.   Concept of robustness.

under the perturbation to the mean value. It is not surprising to note that 'Design B' is a much more robust design than 'Design A' because it is less sensitive to variations.

In Figure 2.13, it is made clear that the robust design point may not be consistent with the global optimal value but rather with the local optimal value. However, since the concept of robustness is concerned with the development of insensitive designs, the local value can be a robust solution when it proves to be the most stable solution in response to variations in the design parameters. It is necessary that the designer ensures the robustness of the solution, which is not sensitive to deviations of the design parameters. In that sense, well-optimised design solutions provide structures that minimise operational costs. These structures perform consistently in the face of unexpected variations throughout their service lifecycle. Therefore, the robustness of structures should be considered during the design parameter stages to reduce the dispersion of structural performance.

2.3.2.2 *Monte Carlo simulation for robustness analysis*

MCS can be used to perform not only reliability analysis but also robustness analysis. As previously discussed in Section 2.3.1.2, this method is an uncertainty propagation method that explicitly shows how uncertainties

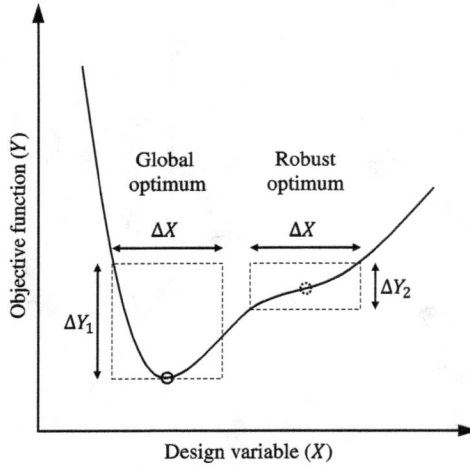

Fig. 2.13. Difference between global optimum and robust optimum.

in input parameters influence the results. If a probability distribution specifies the uncertainties in design parameters, MCS predicts the statistical characteristics of results propagated by the uncertainties. A large number of simulations, considering all uncertain parameters sampled, are conducted to obtain these statistical results. Each simulation result during the MCS process is stored and then put together as a form of probability distribution [22]. Contrary to reliability analysis, which evaluates the limit state function, robustness analysis requires the calculation of the mean and variance of the response function. To assess the robustness, random sampling sets pertaining to input design parameters, N_{sampled}, are created, and these sets are defined by their mean and standard deviation. Then, the sampling sets are simulated and collected to evaluate the response functions. After that, the expected mean values and variances are obtained using equations (2.31) and (2.32), respectively:

$$\mu_X = E[X] = \frac{1}{N_{\text{sampled}}} \sum_{i=1}^{N_{\text{sampled}}} f_i \tag{2.31}$$

$$\sigma_X^2 = V[X] = E[(X - \mu_X)^2] = \frac{1}{N_{\text{sampled}}} \sum_{i=1}^{N_{\text{sampled}}} (f_i - \mu_x)^2 \tag{2.32}$$

where μ_X and σ_X^2 are the mean and standard deviation of the response function f.

In MCS, the greater the number of points sampled, the more accurate the solutions. However, a large number of sampling points aimed at improving the accuracy of solutions leads to substantial computational costs that are necessary to evaluate the response function. This also means that MCS requires numerous simulations when the dimension of the entire design spaces is of large scale.

Chapter 3

Multi-Fidelity Models

As engineering problems have grown increasingly complex and demanding, multi-fidelity models have been more extensively utilised in optimisation over the past three decades. In particular, probabilistic design optimisation requires a large number of computational simulations to evaluate how uncertainties influence outputs. In structural optimisation, these simulations, which are usually based on finite element methods (FEMs) or boundary element methods, are too computationally expensive to be used directly for the entire optimisation process. Models for structural optimisation can be categorised into two types depending on their accuracy: high-fidelity models (HFMs) and low-fidelity models (LFMs) [56]. HFMs, with access to all system information, provide acceptable accuracy but are computationally expensive. LFMs, on the other hand, are computationally economical but less accurate. Thus, LFMs show a certain level of similarity to the response surfaces of HFMs. Multi-fidelity models combine these two models using appropriate methods, relying on the characteristics of problems. The primary aim of multi-fidelity modelling is to offer solutions not only as accurate as those derived from HFMs but also at significantly lower computational costs. Hence, multi-fidelity models involve making trade-offs between solution accuracy and computation time savings.

As shown in Figure 3.1, fidelity may be defined in different ways based on how an HFM can be simplified into an LFM, such as reduced dimensionality, linearisation, partial convergence, simple geometry, simplified physics, and lower refinement [29]. HFM and LFM are generally categorised into three types depending on the nature of each fidelity. Physics describes the differences in how a physical model is presumed

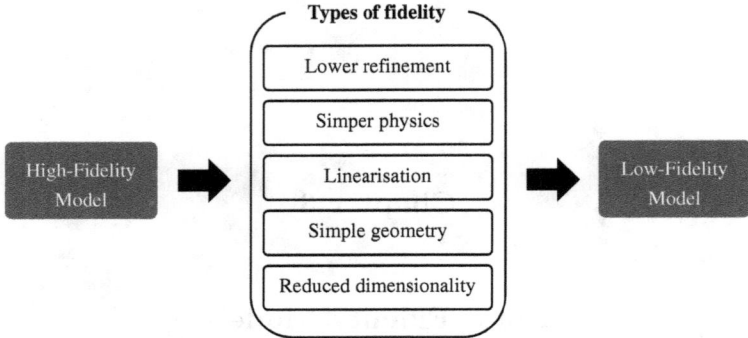

Fig. 3.1.   Examples of fidelity between HFM and LFM.

and implemented. In the beam problem in structural mechanics, for instance, the Euler–Bernoulli beam theory can serve as a physical model representing the LFM, while the Timoshenko beam theory represents the HFM. Numerical accuracy refers to how a defined physical model is calculated using different numerical solvers, such as a linear solver or a nonlinear solver in computational fluid dynamics (CFD). Numerical accuracy also reflects different levels of discretisation or refinement, as seen in FEM models with a fine mesh compared to those with a coarse mesh. Finally, when experiments are carried out as the HFM, computational simulations take the place of the LFM.

Multi-fidelity models constructed by a combination of different fidelity models commonly involve creating surrogate models to obtain greater computational benefits. The surrogate models can be considered approximations with an explicit relationship between design input parameters and output parameters. These models have been widely used to deal with computational challenges by creating a black-box model of a complex system [10]. Multi-fidelity models can be constructed without the surrogate models; however, the computational cost arising from using HFMs and LFMs directly in the optimisation process may still prove to be prohibitively high. The multi-fidelity models based on surrogate models are constructed using deterministic or non-deterministic methods, depending on how parameter estimation is employed to build the surrogate models. Deterministic methods estimate the coefficients of presumed basis functions that minimise the error between sampled design points and the functions. They form surrogate models of the HFM and LFM and compare the response surfaces between them. In comparison, non-deterministic methods consider the uncertainty in either the functions or the coefficients and minimise the uncertainty using the sampled design points. They employ a statistical inference method, such as the Bayesian framework or Gaussian process (GP), to account for the parameter uncertainties without resorting to the computationally expensive standard Monte Carlo simulations (MCSs) [29].

The application area of multi-fidelity models has expanded due to their inherent computational efficiency. The vast majority of the applications involve deterministic optimisation and uncertainty quantification [57]. However, they have been scarcely implemented for probabilistic optimisation considering design uncertainty, except for simple structural design problems covering a small design space. Specifically, multi-fidelity models are utilised in various structural mechanics problems, with the primary fidelities being the structure's dimensionality and the level of discretisation.

This chapter discusses the fundamental theory of surrogate modelling for the construction of multi-fidelity models, including different sampling techniques, surrogate modelling methods, validation methods, and model improvements through sequential design. The probabilistic optimisation of a simple structure is introduced using surrogate models. At the end of this chapter, we discuss the formulation of different multi-fidelity methods, such as response correction methods, space mapping, and autoregressive (AR) modelling.

3.1 Surrogate Models

Computationally intensive design problems in engineering structures incur enormous computational costs. Mitigating this computational cost remains a challenge despite continuous development in computing technologies because the complexity of computer models developed using FEM is also increasing. This high cost can be addressed using high-performance computing. However, design problems based on complex computer models require a large number of calculations, often taking from hours to even days to run a single simulation [26]. Such problems encompass various engineering fields, including structural design optimisation. In particular, reliability and robustness analyses require thousands of MCSs to accurately predict the statistical characteristics. It is not possible to use computer models directly for running the simulations needed if a single simulation requires more than a few hours to run.

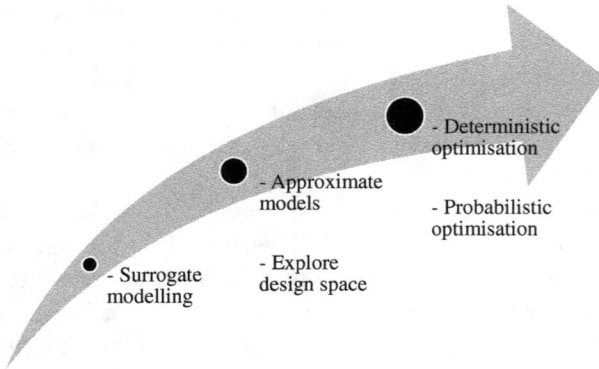

Fig. 3.2.   Application areas of surrogate models.

The surrogate modelling method can improve the overall computational efficiency and achieve acceptable accuracy, when compared with complex computer models. Surrogate models replace computationally expensive models while simultaneously providing an excellent understanding of the relationship between input design parameters and output system performance. The number of input–output data pairs used in the complex computer models and where they are collected in the entire design space are crucial to obtaining an acceptable level of accuracy in the surrogate model and ensuring computational efficiency in its development. Surrogate models can support different engineering areas, as highlighted in Figure 3.2 [28]. They approximate complex computer models across the whole design space to reduce computational costs. Such models enable engineers to explore the design space as well as improve their knowledge of the design problems. Finally, surrogate models can be utilised to find optimal solutions for different types of optimisation problems that are computationally expensive. In particular, surrogate models can be used as an alternative to computationally expensive FEM models in probabilistic applications that demand a vast number of simulations to consider design uncertainties.

There are several common types of surrogate modelling, including artificial neural networks (ANNs), the response surface method (RSM), and GP [13]. The number of design points required to build surrogate models depends on the dimensions of the design space. If a design problem consists of numerous design variables, the computational cost associated with training the surrogate model through FEM simulations also rises dramatically.

3.1.1 *Sampling techniques*

Sampling techniques play a crucial role in creating a surrogate model that accurately represents the original model. If these sampling techniques choose appropriate design points in the design space, the surrogate model is trained accurately, thereby replacing a broad range of characteristics of the original model. The input matrix X denotes the sampled design points that are used to train a surrogate model. The matrix Y comprises the output corresponding to these design points. These two matrices can be expressed as

$$X = \begin{bmatrix} X_1 \\ X_2 \\ \vdots \\ X_n \end{bmatrix} = \begin{bmatrix} x_{11} & x_{12} & \cdots & x_{1m} \\ x_{21} & x_{22} & \cdots & x_{2m} \\ \vdots & \vdots & \ddots & \vdots \\ x_{n1} & x_{n2} & \cdots & x_{nm} \end{bmatrix} \tag{3.1}$$

$$Y = \begin{bmatrix} Y_1 \\ Y_2 \\ \vdots \\ Y_n \end{bmatrix} = \begin{bmatrix} y(X_1) \\ y(X_2) \\ \vdots \\ y(X_n) \end{bmatrix} \tag{3.2}$$

where n and m are the number of sampled design points and the design space dimension, respectively. If the system has one output value from the expensive model, the output Y will be a $(n \times 1)$ vector. If the system has p output values, the output Y will be a $(n \times p)$ matrix.

The quality of a surrogate model relies on how the design points are sampled to set up the input matrix X. Many sampling methods have been developed to explore the entire design space using space-filling approaches. These approaches aim to collect design points in such a way as to ensure they are as evenly distributed as possible while minimising the distance or clustering between different design points. Design of experiments (DOE) is a method adopted to determine the locations of design points that significantly influence the accuracy of the surrogate model. The DOE process is aimed at maximising the amount of information obtained from a limited number of sampling points. Such sampling points estimate performance variability caused by changes in design input parameters. The design matrix, which is built through DOE with input and output data, consists of the values of uncertain design parameters. The most commonly used sampling techniques are simple random sampling, optimal Latin hypercube sampling (OLHS), and Sobol sampling [58]. Each technique has its own advantages and disadvantages in terms of space-filling capability, computational time, and complexity. These methods are introduced and discussed as follows.

3.1.1.1 *Simple random sampling*

Simple random sampling is a basic sampling technique in which the design points to be sampled are selected from a specific design range or a probability distribution. Among the sampling methods, this is the most straightforward. With this sampling method, each design point is selected using a pseudo-random number generator, and every possible design point has the same chance of being sampled. However, due to its inherent randomness, this method can lead to an irregular distribution within the design space, such as clustering of design points and large distances between them. Figure 3.3 illustrates the uneven dispersion of design points caused by simple random sampling. This can result in poor accuracy of the surrogate model because the design points fail to adequately cover the design space. The resulting surrogate model can provide good accuracy in areas with many design points but often produces inaccurate solutions in areas with fewer design points.

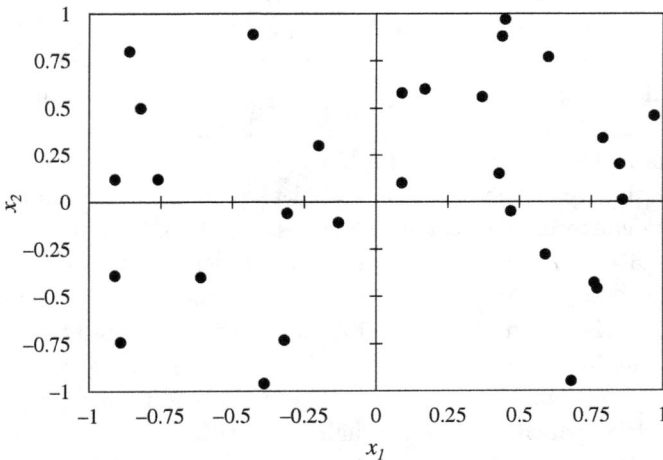

Fig. 3.3.   Distribution of design points using simple random sampling.

3.1.1.2 *Optimal Latin hypercube sampling*

OLHS is a more advanced sampling technique that is derived from Latin hypercube sampling (LHS). LHS is a multivariable sampling method that ensures a non-overlapping design. It is an experimental design that effectively selects the design points across the whole design space. In this method, the distribution for each design variable is divided into an equal probability interval without overlap. Each equally divided probability interval has only one design point. In each interval, design points are randomly selected and rearranged to create different combinations for each point using the available design points. The regularity of uniform probability intervals guarantees that each input design parameter covers every portion of its design space. It reduces both response sensitivities and computational costs associated with sampling design points. The disadvantages of LHS are that the design points are not reproducible since they are combined randomly, and a small number of design points increases the possibility of excluding some areas of the whole design space. OLHS also divides the design space using a uniform probability interval, and these design points in each interval are combined. However, instead of randomly combining the design points like LHS, an optimisation process is applied to an initial matrix created by LHS. A new matrix is obtained by swapping the order of design points in a column of the design matrix. Then, the overall distance between the design points is evaluated. This optimisation process aims to create a matrix in which the design points are distributed as evenly as possible within the design space, defined by the lower and upper ranges. Different kinds of optimality criteria may be employed to achieve such an even distribution of design points. One of these is the max-min distance criterion, which involves maximising the minimum distance between two design points, as expressed in equations (3.3) and (3.4):

$$\max \left(\sum_{1 \leq i,j \leq n, i \neq j} \min d(x_i, x_j) \right) \tag{3.3}$$

where $d(x_i, x_j)$ is the distance between the two different design points x_i and x_j, and n is the required number of design points.

The distance $d(x_i, x_j)$ between two points is defined as

$$d(x_i, x_j) = d_{ij} = \left(\sum_{k=1}^{m} |x_{ik} - x_{jk}|^2 \right)^{1/2} \tag{3.4}$$

where m is the number of design variables.

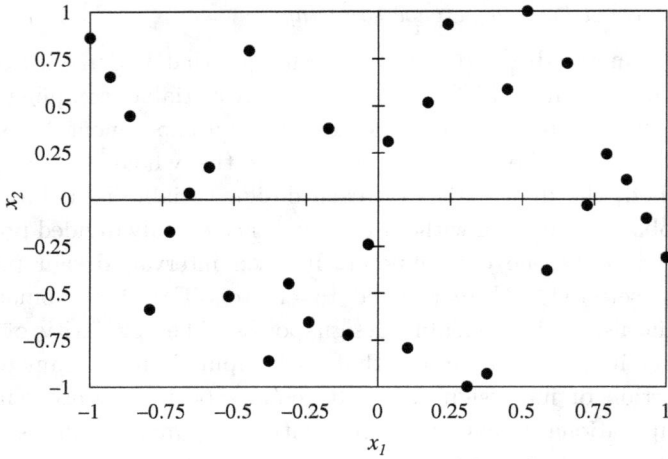

Fig. 3.4.   Distribution of design points using Latin hypercube sampling.

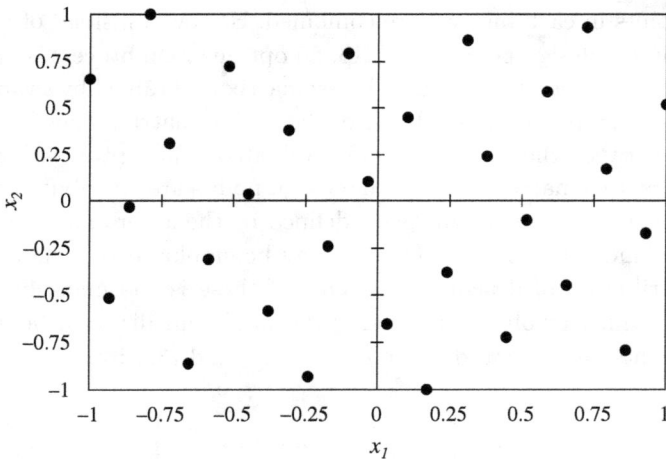

Fig. 3.5.   Distribution of design points using optimal Latin hypercube sampling.

Equation (3.4) evaluates all design points $i, j = 1, 2, \ldots, n$, except $i \neq j$, and delivers the distance matrix d for the optimised design matrix. OLHS provides an excellent opportunity to train the surrogate model due to its outstanding space-filling capability. As illustrated in Figures 3.4 and 3.5,

OLHS collects design points more evenly across the entire design compared to LHS.

3.1.1.3 *Sobol sampling*

Sobol sampling involves a quasi-random sequence that provides more uniformly distributed design points than the simple random sampling and LHS methods. This sampling method selects design points while taking into consideration earlier sampled design points and avoiding clusters and large spaces between them. The method enables the development of high-accuracy surrogate models using a smaller number of design points than simple random sampling and OLHS. A Sobol sequence involves considerably more complex mathematical formulations than other sampling methods but provides more robust results with reduced variability [58].

3.1.2 *Surrogate modelling methods*

Surrogate models are developed with the aim of handling a large number of computational simulations for evaluating objective functions and constraints in a variety of optimisation problems. These models enable us to carry out optimisation problems at an affordable computational cost. Such surrogate models are developed from observations, such as experiments or numerical simulations. They have an explicit mathematical form concerning the relationship between input and output parameters. There are several different methods to construct surrogate models. This section introduces two of the most common methods employed in computer experiments: ANNs and GP.

3.1.2.1 *Artificial neural networks*

An ANN is a network of simple elements known as neurons. These neurons take input data and adjust their activations according to that data. Then, they produce an output associated with the input data and activation. An ANN is made up of three components: neurons, weights, and the activation function [59]. In particular, the radial basis function (RBF), a type of ANN, uses radial units as its activation function. The output of this network is a linear combination of input and neuron parameters. This network architecture comprises three layers: an input layer, a hidden layer consisting of the RBF activation function, and an output layer. The input layer is a vector of real numbers $x \in \mathbb{R}^n$, while the output layer is a scalar function of

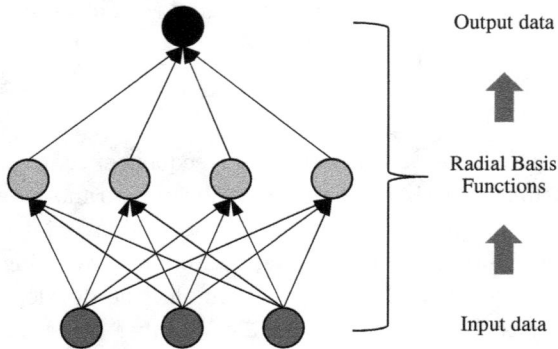

Fig. 3.6.   Basic form of a radial basis function.

the input vector. The basic idea of the RBF is that these functions, which depend on the distance from a central vector, exhibit radial symmetry. One of the advantages of the RBF is that the interpolation problem is not sensitive to the dimension of the design space in which the data points lie. This enables multi-variable functions to be approximated using a linear combination of single-variable functions [60]. All inputs are connected to each hidden layer in the basic form, as shown in Figure 3.6. The input parameter is used as input data to the RBF, and the output data of this network are a linear combination of the outcomes of the RBF.

A basic RBF model is mathematically expressed as

$$h(x) = \sum_{n=1}^{N} w_n \exp\left(-\gamma \|x - x_n\|^2\right) \tag{3.5}$$

Here, N is the number of neurons in the hidden layer. $h(x)$ and w_n are the hypothesis and the weight of neuron n, respectively. $\|x - x_n\|$ is the radial distance between the input and the central point of neuron n, where the norm is typically calculated in terms of the Euclidean distance.

As shown in equation (3.5), each pair (x_n, y_n) in the design space influences $h(x)$ based on the radial distance $\|x - x_i\|$. A learning algorithm finds the weights $w_1, w_2, \ldots, w_n$ from the equation using the training data $D = (x_1, y_1), (x_2, y_2), \ldots, (x_n, y_n)$. These weights minimise the errors between $h(x_n)$ and y_n. Table 3.1 shows that the basis function in equation (3.5), $\exp\left(-\gamma \|x - x_n\|^2\right)$, can be converted into different basis functions ϕ [61].

Table 3.1.   Typical radial basis functions.

Name	Basis function
Linear	$\phi = \|x - x_n\|$
Cubic	$\phi = \|x - x_n\|^3$
Thin-plate spline	$\phi = \|x - x_n\|^2 \ln(\gamma\|x - x_n\|)$
Gaussian	$\phi = \exp(-\gamma\|x - x_n\|^2)$
Multi-quadric	$\phi = \sqrt{\|x - x_n\|^2 + \gamma^2}$
Inverse multi-quadric	$\phi = \dfrac{1}{\sqrt{\|x - x_n\|^2 + \gamma^2}}$

Equation (3.5) consists of N equations in N unknowns; therefore, the solution can be calculated as

$$
\begin{bmatrix}
\exp(-\gamma\|x_1 - x_n\|^2) & \cdots & \exp(-\gamma\|x_1 - x_n\|^2) \\
\vdots & \ddots & \vdots \\
\exp(-\gamma\|x_N - x_n\|^2) & \cdots & \exp(-\gamma\|x_N - x_n\|^2)
\end{bmatrix}
\begin{bmatrix} w_1 \\ \vdots \\ w_N \end{bmatrix}
=
\begin{bmatrix} y_1 \\ \vdots \\ y_N \end{bmatrix}
\tag{3.6}
$$

$$
\begin{bmatrix} w_1 \\ \vdots \\ w_N \end{bmatrix}
=
\left(
\begin{bmatrix}
\exp(-\gamma\|x_1 - x_n\|^2) & \cdots & \exp(-\gamma\|x_1 - x_n\|^2) \\
\vdots & \ddots & \vdots \\
\exp(-\gamma\|x_N - x_n\|^2) & \cdots & \exp(-\gamma\|x_N - x_n\|^2)
\end{bmatrix}
\right)^{-1}
\begin{bmatrix} y_1 \\ \vdots \\ y_N \end{bmatrix}
\tag{3.7}
$$

$$
W = \Phi^{-1} Y
\tag{3.8}
$$

where Φ is the matrix of the basis function.

Once the vector of weights is obtained, the surrogate model created using the RBF can represent a complex computer model and provide outputs concerning different input data that were not used to train the model. As can be seen in equations (3.6) and (3.7), the value of γ in the Gaussian basis function affects the outcomes of the surrogate model. When the value of γ is small, the Gaussian function exhibits a wide distribution, whereas a larger value of γ results in a narrower distribution. Furthermore, γ significantly affects the distribution of design points, depending on their location and the spacing between them. Figures 3.7 and 3.8 display how the value of γ influences interpolation with Gaussian functions [62]. For example, if three points are selected from the red curve in Figure 3.7, the combination of these three interpolations exactly passes through these points. However, the small grey curves represent the contribution according to each interpolation. The red curve is determined by adding up all the

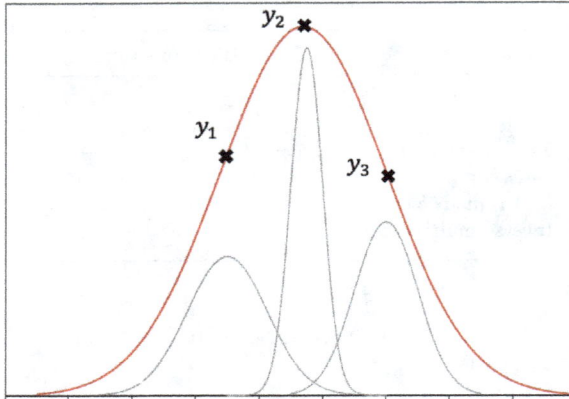

Fig. 3.7.   RBF with a small value of γ.

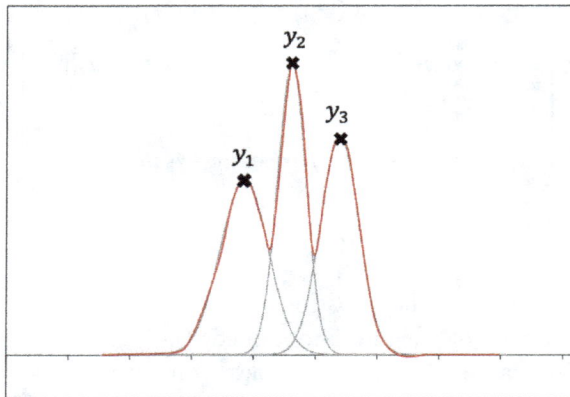

Fig. 3.8.   RBF with a large value of γ.

products obtained by multiplying the weights w_n and their corresponding Gaussian $\exp(-\gamma\|x - x_n\|^2)$. This curve yields the correct output data, i.e., y_1, y_2, and y_3. The width set by a small value of γ ensures successful interpolation, thereby resulting in a reasonable prediction between two different points. In contrast, a large value of γ leads to an incomplete interpolation between different points because each curve falls sharply, although the red curve in Figure 3.8 gives the correct output data at the selected points. However, the interpolation of the three grey curves is insufficient to deliver a satisfactory prediction between different points due to the narrow distribution. The two figures show that the value of γ is important in determining the distance between the points sampled, as this affects the interpolation.

3.1.2.2 *Gaussian process*

GP regression describes a supervised learning problem where both training and testing datasets consist of input–output pairs of observations: $D = (\mathbf{x_1}, \mathbf{y_1}), (\mathbf{x_2}, \mathbf{y_2}), \ldots, (\mathbf{x_n}, \mathbf{y_n})$. GP defines a collection of random variables, a finite number of which have a joint Gaussian distribution [63]. Such a GP is expressed using its mean function and covariance function:

$$\mu(x) = E[f(X)] \tag{3.9}$$

$$k(x, x') = E[(f(x) - \mu(x))((f(x') - \mu(x')))] \tag{3.10}$$

where $\mu(x)$ and $k(x, x')$ are the mean function and the covariance function, respectively. $f(x)$ is a Gaussian process: $f(x) \sim \mathcal{GP}(\mu(x), k(x, x'))$.

The definition of GP necessitates consistency, which is also called the marginalisation property. This property stipulates that if a GP states $(y_1, y_2) \sim N(\mu, \sigma)$, it must also state $y_1 \sim N(\mu_1, \sigma_{11})$, where σ_{11} is the related submatrix of σ. This safeguards the distribution of the subset from being affected by the examination of a greater set of variables. The covariance function $k(x, x')$ is a positive definite kernel that represents the dependence structure of a GP. This function characterises the covariance between different random variables. The covariance matrix $cov(f(x), f(x'))$ based on the squared exponential covariance function is a popular covariance function, which is written as

$$cov(f(x), f(x')) = k(x, x') = \exp\left(-\frac{1}{2}|x - x'|^2\right) \tag{3.11}$$

where $f(x)$ and $f(x')$ denote the GP concerning x and x', respectively, and $|x - x'|$ stands for the Euclidean norm between x and x'.

It should be noted that a function of the input data defines the covariance between the output data of GP regression. The covariance in equation (3.11) is nearly unity for input data points that are in close proximity, while it decreases as the distance between the points increases. If the $|x - x'|$ term in the equation is replaced by $|x - x'|/\theta$ for a positive characteristic length scale θ that can be assumed as the distance moved in the input space before the function value changes considerably, the length scale can act as a parameter to control the oscillation frequencies of the

GP regression model. Also, the process variance σ^2 of the random function can be added to control the range of variation of the model. Finally, the covariance matrix based on the squared exponential covariance function can be expressed by equation (3.12). The length scales, also known as hyperparameters, should be optimised because they are essential to creating an excellent surrogate model using GP. The process of finding these optimal values is called hyperparameter estimation.

$$cov(f(x), f(x')) = \sigma^2 k(\theta, x, x') = \sigma^2 \exp\left(-\frac{1}{2\theta^2}|x - x'|^2\right) \tag{3.12}$$

When the observations are noise-free, $\{(x_i, y_i) \mid i = 1, \ldots, n\}$, the training data provide the function information to be incorporated into the GP regression model. The joint distribution of the training output datasets Y and the test output datasets Y_* based on the prior is expressed as

$$\begin{bmatrix} Y \\ Y_* \end{bmatrix} \sim \mathcal{N}\left(0, \begin{bmatrix} K(X,X) & K(X,X_*) \\ K(X_*,X) & K(X_*,X_*) \end{bmatrix}\right) \tag{3.13}$$

where $K(X,X)$ is the $(n \times n)$ covariance matrix consisting of all the combinations of n training data points and n_* testing data points. Similarly, $K(X,X_*)$, $K(X_*,X)$, and $K(X_*,X_*)$ denote the covariance matrices of their components.

The prior assigns a value to represent the pairwise correlation between different input data points and reflects prior knowledge about the characteristics of the function being estimated. This joint prior distribution should be defined to accommodate the function along with the observed design data points so that the posterior distribution is obtained over the function. This posterior distribution corresponds to conditioning the joint Gaussian prior distribution on the design data points, and it can be given as

$$Y_* \mid X, X_*, Y \sim \mathcal{N}\left(K(X_*,X)K(X,X)^{-1}Y,\right.$$
$$\left. K(X_*,X_*) - K(X_*,X)K(X,X)^{-1}K(X,X_*)\right) \tag{3.14}$$

The new output data Y_*, associated with the testing input data points X_*, can be taken from the joint posterior distribution using the mean and covariance matrices. This conditional distribution provides the predictive equations for the GP regression model as

$$p(Y \mid X, Y, X_*) = \mathcal{N}(Y \quad \mu(X_*), \sigma^2(X_*)) \qquad (3.15)$$

$$\mu(X_*) = K(X_*, X)K(X, X)^{-1}Y \qquad (3.16)$$

$$\sigma^2(X_*) = K(X_*, X_*) - K(X_*, X)K(X, X)^{-1}K(X, X_*)^T \qquad (3.17)$$

Predictions are estimated using the posterior mean μ_*, while the variance associated with these predictions is computed using the posterior variance σ^2.

In order for the GP regression model to predict the response at new testing points X_*, the hyperparameters of the length scale θ and the process variance σ^2 in equation (3.12) must be determined appropriately. Different methods of optimising these parameters, such as maximum likelihood estimation, have been developed [63]. The maximum likelihood estimation used here is a popular method to estimate the parameters. The maximum log-likelihood of the GP regression model can be expressed by

$$\log p(Y|X) = -\frac{1}{2}\log|K| = -\frac{1}{2}Y^T K^{-1}Y - \frac{n}{2}\log 2\pi \qquad (3.18)$$

where K is $K(X, X)$ and n is the number of training data points.

3.1.3 *Validation of surrogate models*

When a surrogate model is constructed using an ANN or GP, it should be validated to ensure that it is not only computationally efficient but also provides accurate solutions. There are two popular validation methods used to evaluate a newly created surrogate model: separation and cross-validation [64]. As shown in Figure 3.9, the separation method requires

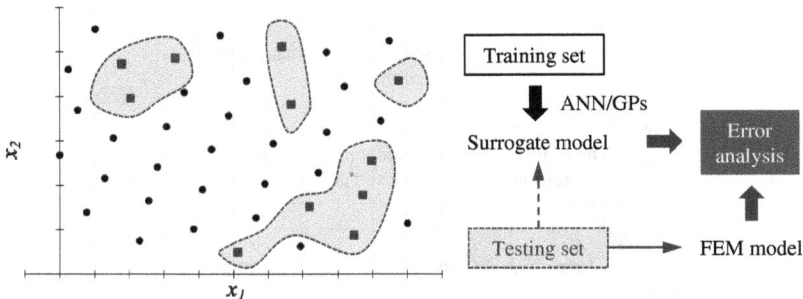

Fig. 3.9. Separation method.

two different datasets, training and testing datasets, which are determined using sampling techniques, such as simple random sampling or OLHS. The training dataset is used to create the surrogate model using global approximation methods, such as an ANN or GP. Once the surrogate model is constructed, the testing dataset is used to assess its accuracy by comparing its output values with those of the original model. The testing dataset is a hidden dataset that is considered only once and should remain inaccessible until the surrogate model is created using the training dataset. Therefore, the testing dataset is not considered during the data training process when creating the surrogate model. In contrast to the separation method, the cross-validation method consists of different phases. Each phase divides a sampled dataset into training and testing datasets. Then, the surrogate model created using the training dataset is validated using the testing dataset. This error analysis is also called the leave-one-out cross-validation method since each phase requires at least one design point in the testing dataset. The more points added in each cross-validation phase, the longer it takes to validate the model. Figure 3.10 shows the general phase of the cross-validation method.

The differences measured by these two validation methods can be quantified through proper error analysis. There are many types of error analysis, such as average, maximum, root-mean-square, and R-squared error analyses. The root-mean-square error analysis used here calculates

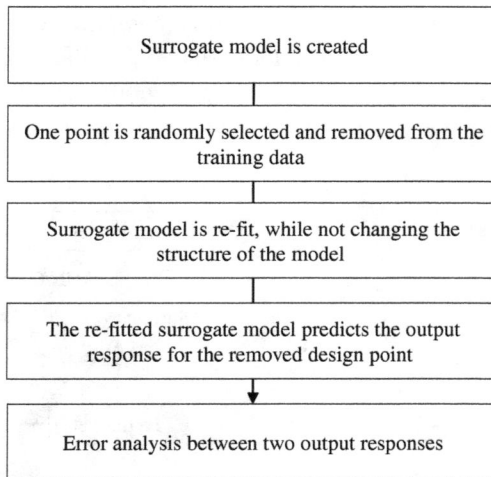

| Surrogate model is created |
| One point is randomly selected and removed from the training data |
| Surrogate model is re-fit, while not changing the structure of the model |
| The re-fitted surrogate model predicts the output response for the removed design point |
| Error analysis between two output responses |

Fig. 3.10.   Cross-validation method.

the squared differences between the actual output values from the original model and the predicted output values from the surrogate model. These squared differences are normalised using the difference between the maximum and minimum actual output values in the design range. The root-mean-squared error is computed as

$$\text{Error} = \frac{\sqrt{\dfrac{\sum_{k=1}^{n}(\text{Actual output} - \text{Predicted output})^2}{n}}}{\text{Maximum actual output} - \text{Minimum actual output}} \qquad (3.19)$$

where n is the number of testing design points.

3.1.4 *Sequential design*

The computational time spent on constructing a surrogate model is dominated by the time required to obtain the actual output values corresponding to the sample design points. A sequential sampling process enables designers to improve the accuracy of approximations as well as computational efficiency. As shown in Figure 3.11, this sampling process aims to add more sampling points to the surrogate modelling process, thereby advancing the quality of the training dataset [65]. To initialise the surrogate modelling process, the sequential sampling method begins with an initial sampling set, which is collected through a sampling technique.

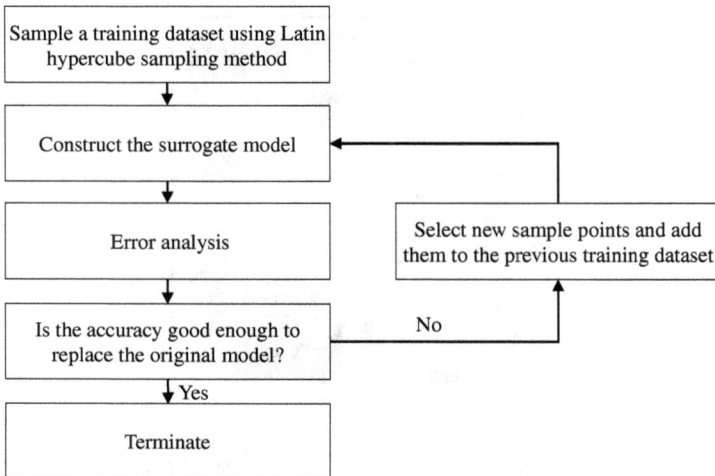

Fig. 3.11.   Sampling procedure for sequential design.

Then, the model selects additional sampling points for this initial set using the max-min scaled distance approach, in which the new sampling points maximise the minimum distance between any two distinct sampling points in the following sampling set. When the accuracy of the surrogate model using the updated training dataset satisfies the required level, the iterative sampling process can be terminated. Unless the process reaches the specified accuracy, it continues to use more new sampling points until it fulfils the tolerance requirement. In this manner, the sequential design ensures that a surrogate model with acceptable accuracy can be created using a more computationally efficient approach than other sampling approaches.

3.1.5 *Numerical example*

In this numerical example, reliability-based design optimisation (RBDO), a type of probabilistic optimisation method, was carried out to design an isotropic steel plate. This example clearly demonstrates how surrogate models are utilised in the probabilistic optimisation process. The isotropic steel drain cover, illustrated in Figure 3.12, was optimised in the RBDO process, considering the uncertainties in the design variables. The objective of the optimisation problem is to minimise both the cost of the drain cover with respect to its mass and the vertical displacement under a

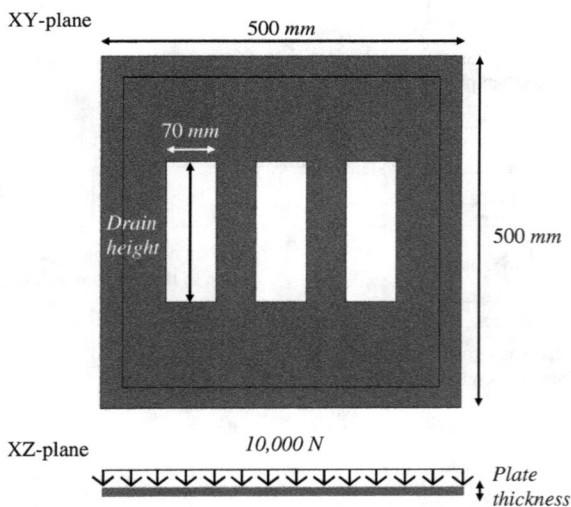

Fig. 3.12.   Isotropic steel drain cover with two design variables.

Table 3.2.   Problem definition.

Description		Value
Material properties	Young's modulus (GPa)	138
	Poisson's ratio	0.3
Dimension	Outer dimension (mm)	500 × 500
	Drain width (mm)	70
Pressure load	Total force distribution (N)	10,000
Design variables	Drain height (mm)	$187 < L < 313$
	Plate thickness (mm)	$3.75 < t < 6.25$
Optimisation	NSGA-II	
Constraint	Maximum von Mises stress (GPa)	160
Objectives	Cost (referring to price per unit mass)	Minimise
	Out-of-plane displacement (mm)	Minimise

uniformly distributed pressure. The maximum stress was defined by a constraint to ensure that the material does not exceed half of its yield point for safety. The height of the drain and the thickness of the steel plate were considered design variables. The FEM model of the drain cover was created using Abaqus and Isight [66, 67]. The problem definition, including dimensions, material properties, loading conditions, and optimisation, is described in Table 3.2. It should be noted that the pressure load is defined as the total force corresponding to the total pressure applied over the entire surface. The reliability analysis, which is an essential part of RBDO, was conducted using the first-order reliability method (FORM) because this method is computationally affordable when used with the FEM model. This demonstrates how the surrogate model achieves significant computational savings compared to using only the FEM model. The gradient of the constraint was calculated using the finite difference method with a step size of 1%. The iteration for the most probable point (MPP) was continued until the reliability index converged to within 0.1%. The design uncertainties in drain height and plate thickness were determined using coefficients of variation, namely, 0.3 and 0.1, respectively. The constraint was 160.0 GPa, which is half of the yield stress of steel.

Figure 3.13 depicts the process of creating the surrogate model for this optimisation problem. After a sufficient number of design points in the design range were sampled through OLHS, the FEM solver calculated the output values using the sampled design points to set up the design matrix. In this example, RBFs with a Gaussian basis function were employed to create the surrogate model. Then, the created surrogate

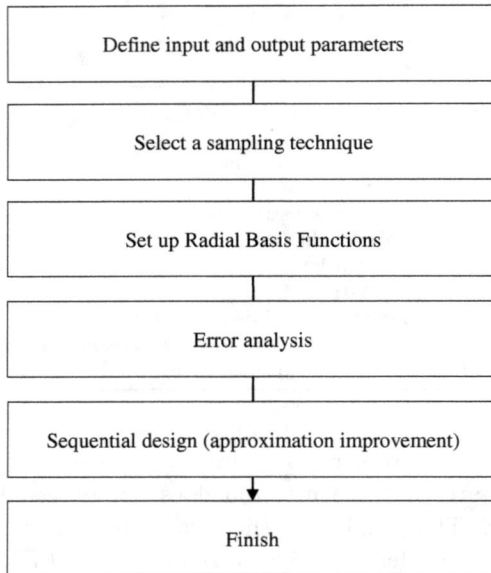

Fig. 3.13.   Surrogate modelling process.

model was evaluated using the testing dataset shown in Table 3.3. The sequential sampling method was also considered to improve the quality of the surrogate model. Table 3.4 presents the RBDO results obtained using FORM and compares them between the FEM and surrogate models. The optimal solutions of each model were selected based on the same vertical displacement on the Pareto front. The costs of the two selected optimal solutions are nearly identical between the FEM model and the surrogate model. The reliability index and the probability of failure also show very similar results. This means that the surrogate model offers trustworthy accuracy in comparison with the FEM model. Most notably, substantial computational gains are achieved through the use of the surrogate model. The optimisation time using the FEM model was over 28 hours, whereas that of the surrogate model was less than an hour. In the surrogate modelling method, the vast majority of the computational cost was caused by the FEM simulation used to create the design matrix from the input and output values. The actual optimisation was considerably faster while providing a nearly identical accuracy level to that of the FEM model.

Table 3.3.   Validation of surrogate model using the separation method.

Testing data point	Drain height (mm)	Plate thickness (mm)	Output	Surrogate model	FEM model	Error analysis (%)
1	190	5.5	Cost (£)	124.8	124.8	0
			Displacement (mm)	2.9	2.9	0.06
			von Mises stress (VMS) (GPa)	117	117.2	0.17
2	160	6	Cost	139.4	139.3	0.03
			Disp.	2.2	2.1	0.94
			VMS	98	97.5	0.17
3	182	4.75	Cost	108.4	108.4	0
			Disp.	4.14	4.14	0
			VMS	146.6	146.8	0.12
4	220	3.8	Cost	84.1	84.2	0.07
			Disp.	6.8	6.7	0.19
			VMS	206.1	193.3	6.64
5	280	6.2	Cost	130.7	130.8	0.11
			Disp.	2.6	2.7	0.92
			VMS	107.2	105.1	2

Table 3.4.   RBDO results comparison between the FEM model and the surrogate model.

	Model	
Parameter	FEM model	Surrogate model
Drain height (mm)	175.6	204.6
Plate thickness (mm)	5.4	5.5
von Mises stress (GPa)	121	120
Cost (£)	122.9	123.2
Vertical displacement (mm)	3	3
Reliability index	2	1.9
Probability of failure	0.04	0.06
Computational time (hh:mm:ss)	28:44:08	0:40:41

3.2 Multi-Fidelity Modelling Methods

The main idea behind multi-fidelity modelling is to create a surrogate model that allows the replacement of HFMs with LFMs through proper mathematical corrections. These correction methods include multiplicative correction, additive correction, comprehensive correction, and space mapping. The first three methods correct the LFM using the difference or ratio of the response surfaces between the HFM and the LFM. Space mapping involves discovering a transformation function of input variables that can be mapped from the high-fidelity space to the low-fidelity space. Conversely, AR modelling calculates the correlation between different fidelity models, which is subsequently used to construct a multi-fidelity model [29, 68].

The construction of a multi-fidelity model requires data from both the HFM and the LFM. The data can be expressed as

$$X_{\mathrm{HF}} = \begin{bmatrix} X_{\mathrm{HF},1} \\ X_{\mathrm{HF},2} \\ \vdots \\ X_{\mathrm{HF},n_{\mathrm{HF}}} \end{bmatrix} = \begin{bmatrix} x_{\mathrm{HF},11} & x_{\mathrm{HF},12} & \cdots & x_{\mathrm{HF},1m} \\ x_{\mathrm{HF},21} & x_{\mathrm{HF},22} & \cdots & x_{\mathrm{HF},2m} \\ \vdots & \vdots & \ddots & \vdots \\ x_{\mathrm{HF},n_{\mathrm{HF}}1} & x_{\mathrm{HF},n_{\mathrm{HF}}2} & \cdots & x_{\mathrm{HF},n_{\mathrm{HF}}m} \end{bmatrix} \tag{3.20}$$

$$X_{\mathrm{LF}} = \begin{bmatrix} X_{\mathrm{LF},1} \\ X_{\mathrm{LF},2} \\ \vdots \\ X_{\mathrm{LF},n_{\mathrm{LF}}} \end{bmatrix} = \begin{bmatrix} x_{\mathrm{LF},11} & x_{\mathrm{LF},12} & \cdots & x_{\mathrm{LF},1m} \\ x_{\mathrm{LF},21} & x_{\mathrm{LF},22} & \cdots & x_{\mathrm{LF},2m} \\ \vdots & \vdots & \ddots & \vdots \\ x_{\mathrm{LF},n_{\mathrm{LF}}1} & x_{\mathrm{LF},n_{\mathrm{LF}}2} & \cdots & x_{\mathrm{LF},n_{\mathrm{LF}}m} \end{bmatrix} \tag{3.21}$$

where n_{HF} and n_{LF} are the number of design points from the HFM and the LFM, respectively, and m is the number of design variables. The output vectors from the HFM and the LFM are written as

$$Y_{\mathrm{HF}} = \begin{bmatrix} Y_{\mathrm{HF},1} \\ Y_{\mathrm{HF},2} \\ \vdots \\ Y_{\mathrm{HF},n_{\mathrm{HF}}} \end{bmatrix} = \begin{bmatrix} Y_{\mathrm{HF}}(X_{\mathrm{HF},1}) \\ Y_{\mathrm{HF}}(X_{\mathrm{HF},2}) \\ \vdots \\ Y_{\mathrm{HF}}(X_{\mathrm{HF},n_{\mathrm{HF}}}) \end{bmatrix} \tag{3.22}$$

$$Y_{\mathrm{LF}} = \begin{bmatrix} Y_{\mathrm{LF},1} \\ Y_{\mathrm{LF},2} \\ \vdots \\ Y_{\mathrm{LF},n_{\mathrm{LF}}} \end{bmatrix} = \begin{bmatrix} Y_{\mathrm{LF}}(X_{\mathrm{LF},1}) \\ Y_{\mathrm{LF}}(X_{\mathrm{LF},2}) \\ \vdots \\ Y_{\mathrm{LF}}(X_{\mathrm{LF},n_{\mathrm{LF}}}) \end{bmatrix} \tag{3.23}$$

where Y_{HF} and Y_{LF} are the outputs from the HFM and the LFM, respectively.

In general, multi-fidelity methods that have been developed so far share the same design space with different fidelity models.

3.2.1 *Response correction methods*

Response correction methods aim to create a surrogate model for the LFM that represents the response surface of the HFM. There are three methods, which vary in how each corrects its LFM in the case where $n_{\mathrm{HF}} = n_{\mathrm{LF}}$ [56]. First, the estimated response of the HFM using multiplicative correction to improve the response of the LFM can be written as

$$\hat{Y}_{\mathrm{HF}}(X) = \beta(X) \cdot Y_{\mathrm{LF}}(X) \tag{3.24}$$

where $\hat{Y}_{\mathrm{HF}}(X)$ is the estimated response of the HFM, $\beta(X)$ is a surrogate model with response ratios between those of the HFM and the LFM, X is a vector of independent design variables, and $Y_{\mathrm{LF}}(X)$ is the response of the LFM.

Second, the estimated response of the HFM using additive correction to correct the response of the LFM can be expressed as

$$\hat{Y}_{\mathrm{HF}}(X) = Y_{\mathrm{LF}}(X) + \delta(X) \tag{3.25}$$

where $\delta(X)$ is a surrogate model of the response differences between the HFM and the LFM.

Finally, the comprehensive methods use both multiplicative and additive corrections. A popular comprehensive method is

$$\hat{Y}_{\mathrm{HF}}(X) = \beta(X) \cdot Y_{\mathrm{LF}}(X) + \delta(X) \tag{3.26}$$

where $\beta(X)$ is the surrogate model of multiplicative correction, and $\delta(X)$ is the surrogate model of additive correction.

Many methods fix the multiplicative correction β as a constant value and employ a surrogate model of additive correction [69]. However, some comprehensive correction methods work with a β that is not constant either. Equation (3.27) displays a new comprehensive correction method that has been proposed using a weighting function based on the traditional correction methods:

$$\hat{Y}_{\text{HF}}(X) = w(X) \cdot \beta(X) \cdot Y_{\text{LF}}(X) + (1 - w(X))(Y_{\text{LF}}(X) + \delta(X)) \quad (3.27)$$

where $w(X)$ is a weighting function.

3.2.2 *Space mapping*

The space mapping method aims to cover different design spaces between the HFM and the LFM so that a small number of high-fidelity simulations are evaluated during the multi-fidelity modelling process [70, 71]. This process continues optimising the multi-fidelity model based on low-fidelity simulations until the stopping criteria of the optimisation are satisfied. For the multi-fidelity model to handle different design spaces, a mapping function between the HFM and the LFM is necessary. Parameter estimation is crucial to establishing the mapping function, which gives an appropriate relationship between two distinct sets of design variables. This method ensures that the low-fidelity design variables can be included as a subset of the high-fidelity ones. Similarly, the high-fidelity design variables can serve as an interpolation of their low-fidelity counterparts. Figure 3.14 illustrates how space mapping constructs a multi-fidelity model. The primary idea is to obtain a suitable model using an appropriate mapping function that offers not only greater computational efficiency than the HFM but

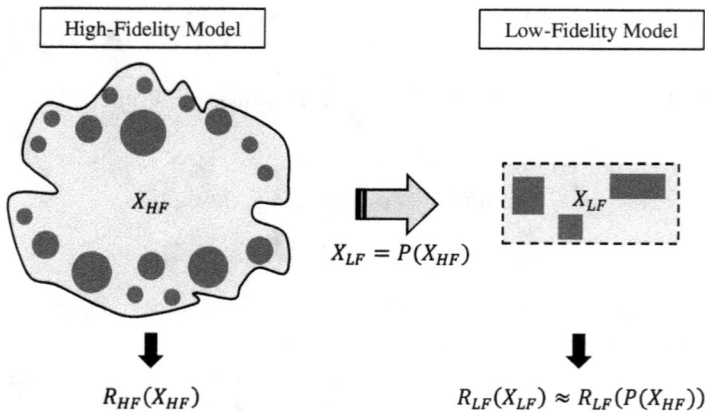

Fig. 3.14.   Space mapping method.

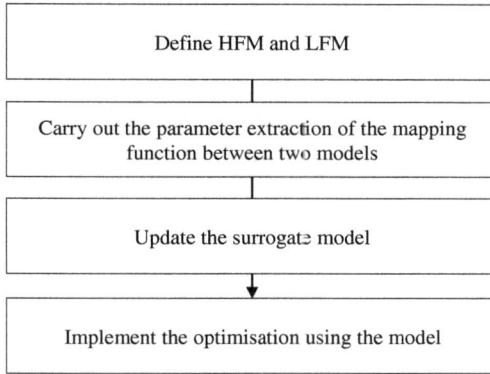

Fig. 3.15.   Procedure of space mapping method.

also, at the very least, improved accuracy compared to the LFM. Space mapping consists of four steps to create the multi-fidelity model, as shown in Figure 3.15.

The space mapping method aims to solve the optimisation problem that is generally given as

$$X^* \triangleq \arg\min_X U(R(X)) \tag{3.28}$$

where $R \in \mathbb{R}^{m \times 1}$ is the vector of m responses of the model, X is the vector of n input variables, and U is an objective function. X^* is the optimal solution to be found.

As shown in Figure 3.14, the design variables of the HFM and the LFM are defined by X_{HF} and $X_{\mathrm{LF}} \in \mathbb{R}^{n \times 1}$, respectively. The corresponding vectors of response are defined by R_{HF} and $R_{\mathrm{LF}} \in \mathbb{R}^{m \times 1}$, respectively. A mapping function P between the HFM and the LFM input variables is expressed by

$$X_{\mathrm{LF}} = P(X_{\mathrm{HF}}) \tag{3.29}$$

Then, the response vectors in a region of interest can be written as

$$R_{\mathrm{LF}}(P(X_{\mathrm{HF}})) = R_{\mathrm{HF}}(X_{\mathrm{HF}}) \tag{3.30}$$

Rather than running a large number of high-fidelity simulations, the mapping function allows us to perform only low-fidelity simulations during the optimisation process. The LFM, having fewer design variables than the HFM, enables the multi-fidelity model to explore the whole design space at an affordable computational cost. At the end of the optimisation process, the model arrives at the optimal solution X_{LF}^* in the low-fidelity design space, which can be transformed into the optimal solution X_{HF}^* in the high-fidelity design space using the inverse mapping function as follows:

$$X_{HF}^* \triangleq P^{-1}(X_{LF}^*) \tag{3.31}$$

Parameter extraction performed to set up the mapping function is a sub-optimisation problem, which is expressed by

$$X_{LF}^{(j)} \triangleq \arg \min_{X_{LF}} \left\| R_{HF}\left(X_{HF}^{(j)}\right) - R_{LF}(X_{LF}) \right\| \tag{3.32}$$

where j is the number of iterations required to find the parameters of the mapping function.

The error associated with the parameter extraction is defined by

$$\epsilon \triangleq \left\| R_{HF}\left(X_{HF}^{(j)}\right) - R_{LF}\left(X_{LF}^{(j)}\right) \right\| = \min_{X_{LF}} \left\| R_{HF}\left(X_{HF}^{(j)}\right) - R_{LF}(X_{LF}) \right\|$$
$$\tag{3.33}$$

The parameters satisfying equations (3.32) and (3.33) enable the HFM's design variables to be transformed into the LFM's design variables. Then, the optimisation problem can be solved in the low-fidelity design space, which requires less computational time to find the optimal solution.

If a mapping function between two different design spaces is linear, the relationship between them can be presumed as

$$X_{LF} = P^{(j)}(X_{HF}) = A^{(j)} X_{HF} + B^{(j)} \tag{3.34}$$

where $A^{(j)} \in \mathbb{R}^{n \times m}$ and $B^{(j)} \in \mathbb{R}^{n \times 1}$.

3.2.3 *Autoregressive modelling*

The AR framework is an advanced GP regression model that constructs multi-fidelity models using different fidelity models [11]. In contrast to response correction methods, this framework can be implemented in cases where $n_{HF} \neq n_{LF}$. The AR regression model utilises the correlation between the information datasets of the most accurate but expensive model, the HFM, and the less accurate but cheaper model, the LFM. Thus, two different information datasets are used in the AR framework: a set of high-fidelity data, (X_{HF}, Y_{HF}), where X_{HF} is an $n_{HF} \times m$ matrix and Y_{HF} is a column vector of length n_{HF}; and a set of low-fidelity data, (X_{LF}, Y_{LF}), where X_{LF} is an $n_{LF} \times m$ matrix and Y_{LF} is a column vector of length n_{LF}.

In this framework, the high-fidelity design points X_{HF} need to be a subset of the low-fidelity design points X_{LF}, $(X_{\text{HF}} \subseteq X_{\text{LF}})$. A more detailed explanation of the mathematical formulations can be found in Refs. [72, 73].

The AR regression model consists of a scaling parameter and GP regression models to approximate the HFM using the LFM. The model can be written as

$$Y_{\text{HF}}(X) = \rho Y_{\text{LF}}(X) + f_{\delta}(X) \tag{3.35}$$

where $f_{\text{HF}}(\cdot)$ and $f_{\text{LF}}(\cdot)$ represent the GP models created using the data from the high-fidelity and low-fidelity datasets, respectively, and ρ is a scaling parameter that quantifies the correlation between the two different fidelity outputs. $f_{\delta}(X)$ is another GP model that represents the difference between $f_{\text{HF}}(X)$ and $\rho f_{\text{LF}}(X)$.

In GP regression modelling, the covariance matrix $cov(f(x), f(x')) = \sigma^2 k(x, x')$ has a simple form. Conversely, in the AR regression framework, the covariance matrix is more complex because of the correlation between different fidelity models. This matrix can be represented as

$$cov(Y_{\text{LF}}, Y_{\text{LF}}) = cov(f_{\text{LF}}(X_{\text{LF}}), f_{\text{LF}}(X_{\text{LF}}))$$
$$= \sigma_{\text{LF}}^2 k(\theta_{\text{LF}}, X_{\text{LF}}, X_{\text{LF}}) \tag{3.36}$$

$$cov(Y_{\text{HF}}, Y_{\text{LF}}) = cov(f_{\text{HF}}(X_{\text{HF}}), f_{\text{LF}}(X_{\text{LF}}))$$
$$= \rho \sigma_{\text{LF}}^2 k(\theta_{\text{LF}}, X_{\text{HF}}, X_{\text{LF}}) \tag{3.37}$$

$$cov(Y_{\text{HF}}, Y_{\text{HF}}) = cov(f_{\text{HF}}(X_{\text{HF}}), f_{\text{HF}}(X_{\text{HF}}))$$
$$= cov(\rho f_{\text{LF}}(X_{\text{HF}}) + f_{\delta}(X_{\text{HF}}), \rho f_{\text{LF}}(X_{\text{HF}}) + f_{\delta}(X_{\text{HF}}))$$
$$= \rho^2 cov(f_{\text{LF}}(X_{\text{HF}}), f_{\text{LF}}(X_{\text{HF}}))$$
$$+ cov(f_{\delta}(X_{\text{HF}}), f_{\delta}(X_{\text{HF}}))$$
$$= \rho^2 \sigma_{\text{LF}}^2 k(\theta_{\text{LF}}, X_{\text{HF}}, X_{\text{HF}})$$
$$+ \sigma_{\delta}^2 k(\theta_{\delta}, X_{\text{HF}}, X_{\text{HF}}) \tag{3.38}$$

Thus, the complete covariance matrix, having a dimension of $(n_{\text{LF}} + n_{\text{HF}}) \times (n_{\text{LF}} + n_{\text{HF}})$, is generated by

$$K = \begin{bmatrix} \sigma_{\text{LF}}^2 k(\theta_{\text{LF}}, X_{\text{LF}}, X_{\text{LF}}) & \rho \sigma_{\text{LF}}^2 k(\theta_{\text{LF}}, X_{\text{LF}}, X_{\text{HF}}) \\ \rho \sigma_{\text{LF}}^2 k(\theta_{\text{LF}}, X_{\text{HF}}, X_{\text{LF}}) & \rho^2 \sigma_{\text{LF}}^2 k(\theta_{\text{LF}}, X_{\text{HF}}, X_{\text{HF}}) + \sigma_{\delta}^2 k(\theta_{\delta}, X_{\text{HF}}, X_{\text{HF}}) \end{bmatrix} \tag{3.39}$$

where σ_{LF}^2 and σ_δ^2 are the process variances and θ_{LF} and θ_δ are the hyperparameters corresponding to the covariance functions for the LFM and the difference data, respectively. These parameters and the scaling parameter ρ can be obtained through maximum log-likelihood estimation.

The predictions $\mu_{*\text{HF}}$ and variances $\sigma_{*\text{HF}}^2$ concerning the test input data points X_* can be taken from the joint posterior distribution using the mean and covariance matrices. This multi-fidelity posterior distribution provides the predictive mean and variance at the high-fidelity level represented by

$$p(f_{\text{HF}}|Y_{\text{HF}}, X_{\text{HF}}, f_{*\text{LF}})) = \mathcal{N}(f_{\text{HF}}|\mu_{*\text{HF}}(X_*), \sigma_{*\text{HF}}^2(X_*)) \tag{3.40}$$

$$\mu_{*\text{HF}}(X_*) = \rho\mu_{*\text{LF}}(X_*) + \mu_\delta + K(X_*, X_{\text{HF}})K^{-1}(X, X)$$
$$[Y_{\text{HF}} - \rho\mu{*}\text{LF}(X_{\text{HF}}) - \mu_\delta] \tag{3.41}$$

$$\sigma_{*\text{HF}}^2(X_*) = \rho^2\sigma_{*\text{LF}}^2(X_*) + K(X_*, X_*) - K(X_*, X_{\text{HF}})$$
$$K^{-1}K(X_*, X_{\text{HF}})^T \tag{3.42}$$

where the superscript $(')$ represents the estimated hyperparameters and $K(X_*, X_{\text{HF}})$ is a column vector of length $(n_{\text{LF}} + n_{\text{HF}})$:

$$K(X_*, X_{\text{HF}}) = \begin{bmatrix} \rho'(\sigma_{\text{LF}}')^2 k(\theta_{\text{LF}}', X_{\text{LF}}, X_*) \\ (\rho')^2(\sigma_{\text{LF}}')^2 k(\theta_{\text{LF}}', X_{\text{HF}}, X_*) + (\sigma_\delta')^2 k(\theta_\delta', X_{\text{HF}}, X_*) \end{bmatrix}$$
$$\tag{3.43}$$

Chapter 4

Multi-Fidelity Reliability-Based
Design Optimisation

Reliability-based design optimisation (RBDO) provides many advantages because it enables engineers to consider uncertainties in different design variables used in the design and manufacture of composite structures. The conventional design approach for composite structure design may result in a conservative final design that involves using safety factors to prevent structural failure. In contrast to this design approach, RBDO allows engineers to understand how design uncertainties affect structural reliability or the probability of failure. At the same time, they ensure not only the most critical design areas but also improvements in the reliability of the structures. RBDO includes reliability assessment based on either statistical methods or non-statistical methods. Monte Carlo simulation (MCS) is a statistical method that evaluates the limit state function directly during the optimisation process. The first-order reliability method (FORM) and the second-order reliability method (SORM) calculate reliability using first-order and second-order Taylor series expansions, respectively, to approximate the limit state function. MCS generally offers the most accurate solutions among the three methods since it uses the limit state function directly without approximation. SORM, which uses a higher-order approximation, is more accurate than FORM when the limit state function is particularly highly nonlinear. More details of these reliability methods are introduced in many books and works of literature [8, 9].

There have been many research works on the development and application of RBDO in structural optimisation, such as in isotropic and composite structures. The reliability analysis, which is an essential part of RBDO, has been actively considered in various structural design problems [50, 74]. One remarkable example is the reliability analysis of a 2D rectangular isotropic plate with a central hole subjected to uniaxial tension [50]. The reliability indexes, which represent the probability of success, were compared using different reliability methods, such as MCS, FORM, and SORM. The comparison of the results showed that the two approximation methods obtained similar results to those obtained through MCS. The optimisation process, combined with the reliability analysis, i.e., RBDO, has been widely used in the design of composite structures [24, 75]. One notable example demonstrated an advanced RBDO process for a highly complex composite structure: a multi-stiffened stringer composite panel [15]. This optimisation process found the stacking sequence of the composite panel to maximise the ultimate load in the post-buckling regime. The hybrid mean value algorithm, using the first-order approximation to locate the most probable failure point (MPP), was used. This algorithm proposed a subsequent upcoming point during the reliability analysis depending on the type of the limit state function: convex or concave. Based on the initial stacking sequence provided by DO, the reliability analysis evaluated the reliability of the composite panel associated with uncertainties in material properties.

In general, RBDO is computationally expensive as it accounts for design uncertainties at every design point to evaluate if each design violates the limit state function. In general, the reliability analysis using MCS requires a significant number of high-fidelity finite element method (FEM) simulations, typically exceeding hundreds or even thousands, until the results converge. Even though a single high-fidelity FEM simulation requires only a few seconds to run, the entire optimisation may take several days or even weeks to run the total number of FEM simulations. Surrogate modelling, also called metamodelling, has been garnering attention to address the substantial computational costs incurred when assessing design uncertainties in probabilistic design approaches. Wang *et al.* introduced surrogate modelling approaches as essential tools to perform optimisation. In particular, they showed that surrogate models play a significant role in multi-objective optimisation (MOO). These models improve the computational performance and can clarify our understanding of the effects of design variables. There are many types of surrogate models that have been exploited concerning RBDO [10]. Hassanien *et al.* applied a surrogate model created using the response surface method (RSM) to conduct the reliability analysis of dented pipes [74]. Scarth *et al.* created a surrogate model using support vector machines with Gaussian processes (GPs) to drive the RBDO of a composite aircraft structure [76]. Bacarreza *et al.* used a radial basis function (RBF) to create a surrogate model of a composite stiffened panel [4]. Sampling techniques contribute significantly to the performance of the surrogate model since an inappropriate sampling process might increase the computational costs. The sequential sampling method was developed to offer better sampling performance than other techniques [65, 77].

Although the use of surrogate models reduces computation time, composite structures may require several hours to simulate even a single FEM model, depending on the characteristics of the problem. The concept of multi-fidelity models has been introduced in structural optimisation to address this computational challenge. Multi-fidelity models, which are created using both high-fidelity models (HFMs) and low-fidelity models (LFMs), provide results comparable in accuracy to surrogate models that are solely based on HFMs while providing a notable reduction in computational costs. Extensive research has been carried out with the aim of applying multi-fidelity models to structural optimisation. The vast majority of such works have been confined to the area of DO, which does not consider the uncertainty of design variables. Vitali *et al.* introduced the concept of multi-fidelity models to crack propagation in a composite structure [56]. The idea was to use the ratio and the difference between an LFM and an HFM. Alexandrov *et al.* combined an HFM and an LFM using the multiplicative correction function [78]. This function is constructed through Taylor series approximation, and it makes the LFM follow the response of the HFM. The optimisation of buckling analysis for a laminated shell was conducted using multi-fidelity models implementing an LFM [79]. Response correction surfaces, which are ratios between the HFM and the LFM, were built using various polynomial functions to create the multi-fidelity models. The LFM was changed to a high-order polynomial response surface with improved accuracy and computation time savings through this approximation. Due to its many advantages, multi-fidelity models have widely replaced the use of FEM simulations in different DO problems.

However, its application in the probabilistic optimisation of composite design has received little attention. Prior to the work presented in this monograph, no previous research has been conducted on the topic of RBDO of composite structures using multi-fidelity models. The main objective of the work introduced in this chapter was to develop, for the first time, a multi-fidelity formulation for reliability analysis and RBDO of composite structures. The developed multi-fidelity formulation considers the effects of design uncertainties using different reliability methods. In particular, this work showed that the multi-fidelity formulation can be applied to approximate the limit state function using FORM and SORM while providing nearly identical results to MCS. The multi-fidelity formulation, incorporating the RBDO process, constructs two types of multi-fidelity models depending on how the FEM models are utilised, namely direct and indirect multi-fidelity models. The formulation provides not only similar accuracy to conventional high-fidelity surrogate modelling but also considerable computation time savings. For the first time, the proposed multi-fidelity framework was demonstrated through the reliability analysis and RBDO of a mono-stringer stiffened composite panel to evaluate the computational gains achieved.

This chapter begins by introducing the mathematical formulation of the multi-fidelity method used in this work. It then describes how the multi-fidelity formulation implements both structural reliability analysis and the RBDO framework. At the end of this chapter, two engineering examples of increasing complexity demonstrate the multi-fidelity modelling framework to showcase its potential in handling large-scale problems. The accuracy and computational time savings are evaluated and compared with the results of conventional surrogate models that use only HFMs. The work presented in this chapter is based on the research carried out by the authors [32].

4.1 Multi-Fidelity Models in the Same Design Space

As described in the previous chapter, the response correction methods require the same training data points between the HFM and LFM to construct multi-fidelity models. The primary feature of these methods is that they share the same dimension of the design spaces. This enables an accurate response correction function to be created using the information derived from the same design points of the two different fidelity models. The response correction methods allow the inaccurate LFM to represent the response surface of the accurate HFM using the same design points collected through a proper sampling technique. In this work, the multiplicative and additive correction functions were used [56]. Two types of multi-fidelity models, depending on how the LFM works with the multi-fidelity method, are constructed using the response correction methods, namely direct and indirect multi-fidelity models.

In this multi-fidelity modelling method, both the HFM and the LFM require the same number of design points, $n_{\mathrm{HF}} = n_{\mathrm{LF}}$, and share the same dimension of the design spaces with each other. This enables the response correction methods to build a proper response correction surface using each information coming from the HFM and the LFM. Two matrixes representing the ratio and the difference can be created, depending on the number of responses, using the information. These two matrixes can build a surrogate model each for the ratio and the difference. These surrogate models offer the appropriate corrections corresponding to different design points so that the accuracy of the LFM is sufficiently improved to represent the HFM. These models can be created as artificial neural networks (ANNs), as described in Section 3.1.2.1.

The direct type of multi-fidelity model requires the surrogate model of a response correction surface to be chosen. This type involves directly calling the response of the LFM during the multi-fidelity modelling process. Two different multi-fidelity models can be created using these two surrogate models as follows:

$$\hat{Y}_{\mathrm{MF1}}(X) = \beta(X) \cdot Y_{\mathrm{LF}}(X) \tag{4.1}$$

$$\hat{Y}_{\mathrm{MF2}}(X) = Y_{\mathrm{LF}}(X) + \delta(X) \tag{4.2}$$

where X denotes x design points for training with input design parameters. MF1 and MF2 denote the direct multi-fidelity models using different response correction functions. $\beta(X)$ and $\delta(X)$ are the surrogate models of the ratio and difference between the HFM and the LFM, respectively. $Y_{\mathrm{LF}}(X)$ is the response directly calculated using the low-fidelity FEM model.

The indirect type of multi-fidelity model uses two different surrogate models of the LFM and a response correction surface. The indirect type involves approximating the LFM using an ANN. This enables $(Y_{\mathrm{LF}}(X))$ to be replaced by a surrogate model of the LFM, thereby reducing computation time, even if the LFM itself is computationally expensive. Another two different multi-fidelity models are expressed as follows:

$$\hat{Y}_{\mathrm{MF3}}(X) = \beta(X) \cdot \hat{Y}_{\mathrm{LF}}(X) \tag{4.3}$$

$$\hat{Y}_{\mathrm{MF4}}(X) = \hat{Y}_{\mathrm{LF}}(X) + \delta(X) \tag{4.4}$$

where $\hat{Y}_{\mathrm{LF}}(X)$ is the surrogate model of the LFM. MF3 and MF4 denote the indirect multi-fidelity models. $\beta(X)$ and $\delta(X)$ are the surrogate models of the ratio and difference between the HFM and the LFM, respectively.

4.2 Multi-Fidelity Reliability Analysis

A reliability analysis aims to evaluate a limit state function as a constraint and calculate the reliability of a structure. The probability of failure refers to how the structure exhibits reliability under a specific restricted condition. In general, this analysis ensures that powerful computing resources are provided because it needs to calculate the influence of uncertainties in design variables. The multi-fidelity modelling-based reliability analysis developed in this work defines the limit state function. This can be represented as

$$g_{\mathrm{MF}}(X) = R_{\mathrm{MF}}(X) - S_{\mathrm{MF}}(X) \tag{4.5}$$

$$P_{f,\mathrm{MF}} = P[g_{\mathrm{MF}}(Z) < 0] \tag{4.6}$$

where $g_{\mathrm{MF}}(X)$ is the limit state function using the multi-fidelity models, $P_{f,\mathrm{MF}}$ is the probability of failure using the multi-fidelity models, X is a vector of all design variables under consideration, and $R_{\mathrm{MF}}(X)$ and $S_{\mathrm{MF}}(X)$ are the resistance and loading of the structure, respectively, which come from the multi-fidelity models.

If the value of $g_{\mathrm{MF}}(X)$ is less than zero, the structure is not in the safe region. If the value of $g_{\mathrm{MF}}(X)$ equals zero or is more than zero, the structure is in the failure surface or the safe area, respectively. In this work, MCS, FORM, and SORM calculate the probability of failure using the multi-fidelity models blending two different fidelity models.

First, MCS using the multi-fidelity models predicts the probability of failure after $N_{\mathrm{total,MF}}$ multi-fidelity simulations are carried out. This can be expressed by

$$P_{f,\mathrm{MCS,MF}} = \frac{N_{f,\mathrm{MF}}}{N_{\mathrm{total,MF}}} \tag{4.7}$$

where $N_{f,\mathrm{MF}}$ and $N_{\mathrm{total,MF}}$ are the number of failure and the total multi-fidelity simulations conducted. $P_{f,\mathrm{MCS,MF}}$ is the probability of failure using the MCS predicted by the multi-fidelity models.

When MCS estimates the statistical characteristics for reliability analysis, Sobol sampling is used to collect the design points that consider the influence caused by design uncertainties. As described in Section 3.1.1.3, Sobol sampling provides more uniformly distributed design points than simple random sampling since this sampling technique aims to reduce the variance of statistical predictions for MCS. In particular, Sobol sampling offers more robust statistical results than even Latin hypercube sampling for reliability analysis [58].

In contrast to MCS, FORM and SORM approximate the limit state function using the first- and second-order Taylor expansion series to compute the probability of failure, respectively. The calculation of derivatives of the limit state function is essential to determine the accuracy of the solution and the computational costs required to predict a reliability index. There are two typical methods for calculating the derivatives of the limit state function: the finite difference method (FDM) and the implicit differentiation method (IDM) [50]. In this work, the FDM is applied to calculate the probability of failure.

The principle of the FDM is that derivatives in the partial differential equation are approximated by the linear combination of response values defined by a finite grid [80]. The FDM is a relatively challenging method because the sensitivities may be hugely susceptible to the step size chosen. This implies that rounding or truncation errors become considerable, depending on the step size. However, the FDM provides a distinct advantage in calculating derivatives for complex problems, for which analytical solutions do not exist.

There are three standard methods to evaluate the first-order derivatives using the FDM: the forward difference, the backward difference, and the central difference schemes. The first-order forward finite difference scheme is expressed by

$$\left(\frac{\partial S_{\mathrm{MF}}(X)}{\partial X} \right)_i = \frac{S_{\mathrm{MF}}(X_i + \Delta X_i, X_{-i}) - S_{\mathrm{MF}}(X)}{\Delta X_i} \tag{4.8}$$

The first-order backward finite difference scheme is expressed by

$$\left(\frac{\partial S_{\mathrm{MF}}(X)}{\partial X} \right)_i = \frac{S_{\mathrm{MF}}(X) - S_{\mathrm{MF}}(X_i - \Delta X_i, X_{-i})}{\Delta X_i} \tag{4.9}$$

The first-order central finite difference scheme is expressed by

$$\left(\frac{\partial S_{\mathrm{MF}}(X)}{\partial X} \right)_i = \frac{S_{\mathrm{MF}}(X_i + \Delta X_i, X_{-i}) - S_{\mathrm{MF}}(X_i - \Delta X_i, X_{-i})}{2\Delta X_i} \tag{4.10}$$

The second-order finite difference scheme determines the derivatives using the central scheme as follows:

$$\left(\frac{\partial^2 S_{\mathrm{MF}}(X)}{\partial X^2} \right)_i$$
$$= \frac{S_{\mathrm{MF}}(X_i + \Delta X_i, X_{-i}) - 2S_{\mathrm{MF}}(X) + S_{\mathrm{MF}}(X_i - \Delta X_i, X_{-i})}{(\Delta X_i)^2} \tag{4.11}$$

where ΔX_i is the step size and X_{-i} is the vector of the design variable eliminating the variable X_i. The magnitude of the step size ΔX_i affects the accuracy of the derivatives of $S_{\mathrm{MF}}(X)$. A small step size will cause substantial rounding errors, whereas a large step size will cause substantial truncation errors. It was found that the proper value of step size ranges between 0.01% and 0.1% using sensitivity analysis. The step size in this range yielded accurate derivatives of $S_{\mathrm{MF}}(X)$; hence, a value from the range was selected as the step size in this work.

FORM calculates the reliability indexes presenting the shortest distance from the origin to the failure surface in the standard normal distribution. This method transforms the mean value point of the original space (X-space) to the origin of normal space (U-space). This transformation enables the Taylor expansion point to be moved from the mean value point to the most probable failure point (MPP). The first-order Taylor series of the expansion of $g_{\mathrm{MF}}(U)$ at MPP U^* using the multi-fidelity models is defined as

$$\hat{g}_{\mathrm{MF}}(U) = g_{\mathrm{MF}}(U^*) + \sum_{i=1}^{n} \frac{\partial g_{\mathrm{MF}}(U^*)}{\partial U_i}(U_i - U_i^*) \qquad (4.12)$$

The shortest distance, also known as the reliability index, from the origin to the failure surface using the multi-fidelity models is written as

$$\beta_{\mathrm{MF}} = \frac{g_{\mathrm{MF}}(U^*) - \sum_{i=1}^{n} \frac{\partial g_{\mathrm{MF}}(U^*)}{\partial X_i} \sigma_{U_i} U_i^*}{\sqrt{\sum_{i=1}^{n} \left(\frac{\partial g_{\mathrm{MF}}(U^*)}{\partial X_i} \sigma_{X_i}\right)^2}} \qquad (4.13)$$

This process is iterated until the estimate of the reliability index β_{MF} using the multi-fidelity models converges within specific tolerance criteria. When the limit state function is the normal distribution function, the probability of failure using the multi-fidelity models is defined as

$$P_{f,\mathrm{FORM,MF}} = 1 - \Phi(\beta_{\mathrm{MF}}) = \Phi(-\beta_{\mathrm{MF}}) \qquad (4.14)$$

where $P_{f,\mathrm{FORM,MF}}$ is the probability of failure using FORM predicted by the multi-fidelity models and $\Phi(\cdot)$ is the standard normal cumulative distribution function.

Compared to FORM, SORM requires the second-order derivatives to approximate the failure surface to yield more accurate results under the highly nonlinear failure surface. The second-order approximation of the limit state function $g_{MF}(U) = 0$ is derived from the second-order Taylor series expansion at the MPP using the multi-fidelity models:

$$\hat{g}_{MF}(U) = g_{MF}(U^*) + \nabla g_{MF}(U^*)^T (U - U^*)$$
$$+ \frac{1}{2}(U - U^*)^T \nabla^2 g_{MF}(U^*)(U - U^*) \qquad (4.15)$$

where $\nabla^2 g_{MF}(U^*)$ is the Hessian matrix, which is the symmetric matrix of the second derivative of the limit state function.

The Hessian matrix creates additional computational costs during reliability analysis; however, it ensures that SORM provides a more accurate reliability index for a nonlinear limit state function.

The probability of failure using the multi-fidelity models, as predicted by SORM, is defined as

$$P_{f,SORM,MF} = \Phi(-\beta_{MF}) \prod_{j=1}^{n-1} (1 + k_j \beta_{MF})^{-\frac{1}{2}} \qquad (4.16)$$

where $P_{f,SORM,MF}$ is the probability of failure using SORM predicted by the multi-fidelity models and k_j indicates the curvature of the response surface at the MPP.

4.3 Multi-Fidelity Reliability-Based Design Optimisation

The conventional form of RBDO is introduced in equation (2.4). This RBDO process requires computationally intensive models or a surrogate model using many high-fidelity design points. A novel multi-fidelity modelling-based RBDO framework to address the computational challenge stemming from the traditional optimisation process is developed in this work, as illustrated in Figure 4.1. This framework can be divided into two stages: the multi-fidelity modelling process and the multi-fidelity RBDO process.

Before the RBDO process begins, the multi-fidelity modelling process should be conducted to construct multi-fidelity models with adequate quality. The modelling process defines the HFM and LFM based on the discretisation level of the FEM model, which is commonly used in structural optimisation. The type of fidelity can be determined by relying on the characteristics of optimisation problems. Then, the sampling technique — which is the optimal Latin hypercube sampling (OLHS) method in this work — collects appropriate design points to set up the design matrix that

Sample design points	Design of experiments (i.e., OLHS)
FEM calculation to obtain input/output datasets	$y_{HF}^{FEM}(X), y_{LF}^{FEM}(X), \beta(X), \delta(X)$
Build surrogate models using ANNs	$y_{LF}^{ANN}(X), \delta^{ANN}(X), \beta^{ANN}(X)$
Direct multi-fidelity models	$\beta^{ANN}(X) \cdot Y_{LF}^{FEM}(X)$ $\delta^{ANN}(X) + Y_{LF}^{FEM}(X)$
Indirect multi-fidelity models	$\beta^{ANN}(X) \cdot Y_{LF}^{ANN}(X)$ $\delta^{ANN}(X) + Y_{LF}^{ANN}(X)$

Multi-fidelity modelling

Select an optimisation method	Evolutionary methods (i.e., NSGA-II)
Generate new populations	Generation and population numbers
Multi-fidelity modelling-based reliability analysis (MCS, FORM, SORM)	$P_{f,MCS,MF}$ $P_{f,FORM,MF}$ $P_{f,SORM,MF}$
Check stopping criteria	Reliability index convergence
Find the optimal design	Compare accuracy and computational cost with the conventional method

No / Yes

Reliability-Based Design Optimisation

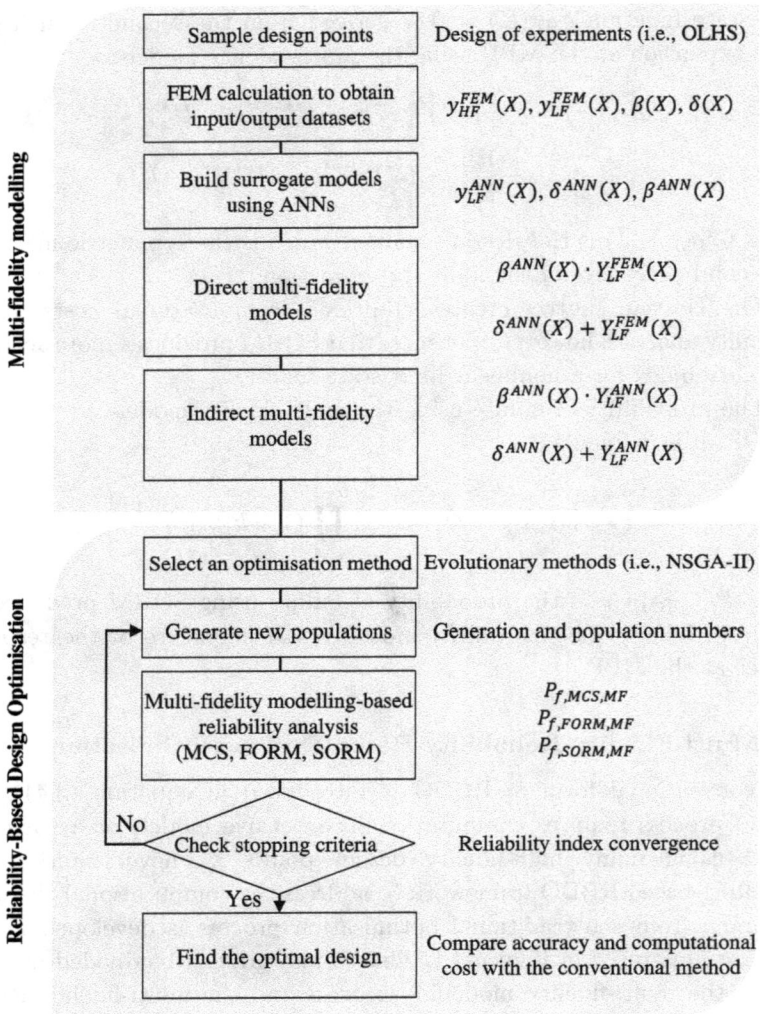

Fig. 4.1. Multi-fidelity modelling-based reliability-based design optimisation framework.

includes all input design parameters in the entire design space. The collected design points are employed to set up the input and output datasets for the two different fidelity models using an FEM solver. An ANN based on RBF using these datasets creates three surrogate models of the LFM, the ratio and the difference between the two fidelity models. The three surrogate models should be evaluated using proper error analysis, such as root-mean-square error analysis, to demonstrate the specified level of accuracy in this framework. Once the models are validated, the multi-fidelity models are constructed using the combination of these three surrogate models and one low-fidelity FEM model, which depends on whether the direct or indirect multi-fidelity modelling approach is adopted. The direct multi-fidelity model considers the response of the composite structures directly from the low-fidelity FEM models during the multi-fidelity modelling process. Conversely, the indirect type calls for a response from the surrogate model using the low-fidelity training datasets. Both models exploit the surrogate models of response correction surfaces so that the LFM represents the nature of the high-fidelity response.

After this modelling process, the RBDO process begins to find the optimal design while exploring the whole design space using the constructed multi-fidelity models. There are many optimisation methods to conduct the RBDO process, and a proper optimisation method can be determined according to the problem characteristics. In particular, evolutionary methods, described in Section 2.1.3, work well with the probabilistic design optimisation of composite structures because these methods are suited to catch the global reliable solution in the nonlinear solution space [44]. When a proper optimisation method is chosen, three different reliability analysis methods, namely MCS, FORM, and SORM, predict the probability of failure for each design point offered by the optimisation method. When the specified stopping criteria are satisfied, the optimal solutions, which are found using the multi-fidelity models, are evaluated in terms of accuracy and computation time savings compared to the conventional RBDO process using high-fidelity modelling-based surrogate models.

4.4 Numerical Examples

This section demonstrates the developed multi-fidelity models using two numerical examples concerning the design of composite structures. A brief outline of each numerical example is described as follows:

- *Numerical Example 1*: Multi-fidelity modelling-based reliability analysis. This numerical example, based on a previous work by the authors [32], aims to demonstrate how multi-fidelity modelling approaches can be used to implement different reliability methods, such as MCS, FORM, and SORM, while increasing the efficiency of conventional high-fidelity reliability analysis for a composite structure.

- *Numerical Example 2*: Multi-fidelity reliability-based design optimisation. This numerical example, also based on the authors' previous work [32], aims to show, for the first time, the potential of utilising multi-fidelity modelling approaches in the RBDO of composite structures. The efficiency as a result of using multi-fidelity models is evaluated as well.

4.4.1 *Model description*

The composite structure considered in this work is a mono-stiffened stringer composite panel, as shown in Figure 4.1. The geometry of this composite panel is parameterised by $X1$, $X2$, $X3$, and $X4$, which represent the stringer foot length, the stringer height, the horizontal distance between the top and foot of the stringer, and the stringer top length, respectively. The material properties and dimensions of the panel are denoted in Table 4.1. The composite panel consists of the skin and the stringer. It is clamped at both the right-hand and left-hand ends, but the left-hand end is free to move in the loading direction (z-direction in Figure 4.2). There are no constraints on either side of the skin. The composite panel was analysed in the linear-buckling regime to obtain the maximum first buckling load. In the example for reliability analysis, linear buckling was considered when the applied load has eccentricity regarding the centre of mass of the stringer (y-direction in the figure). In the example of RBDO, the load was applied at the centre of mass without eccentricity. In these two examples, an HFM and an LFM should be defined first to create the multi-fidelity models. The level of discretisation of FEM models was considered as a type of fidelity in this work. The FEM models are built and calculated using Abaqus/CAE. The mesh convergence study was conducted to select both the HFM and

Table 4.1. Material properties and dimensions of the composite structure.

Parameter	Notation	Value
Longitudinal modulus of elasticity (GPa)	E_{11}	139
Transversal modulus of elasticity (GPa)	$E_{22} = E_{33}$	8.1
Poisson's ratio	ν	0.33
Out-of-plane shear modulus (GPa)	$G_{12} = G_{13}$	3.1
In-plane shear modulus (GPa)	G_{23}	4.8
Skin and stringer thickness (mm)	t	2.208
Skin and stringer layup (degree)		$[45/-45/0/0/90/0]_{sym}$
Panel length (mm)	L	600
Panel width (mm)	W	250

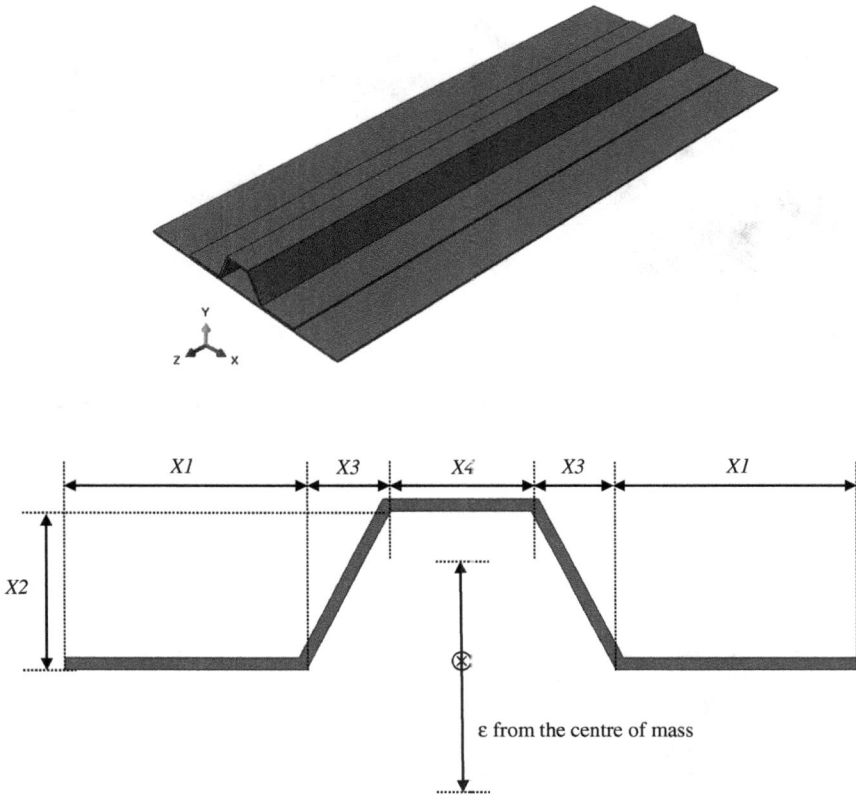

Fig. 4.2. Mono-stringer stiffened composite structure with four geometry parameters.

the LFM with different accuracies and computational costs. As shown in Figure 4.3, the mesh size was determined to be 4.0 and 30.0 mm for the HFM and LFM, respectively. The accuracy difference between the two models was 10%, and the LFM was 80% more efficient than the HFM in terms of computation time.

4.4.2 *Multi-fidelity reliability analysis*

The multi-fidelity reliability analyses using MCS, FORM, and SORM were performed on the mono-stringer stiffened composite panel. The composite panel was loaded by a compressive axial load with eccentricity. Five random variables, four parameters defining the geometry. and one load eccentricity were considered. The limit state function, which is made up of five random

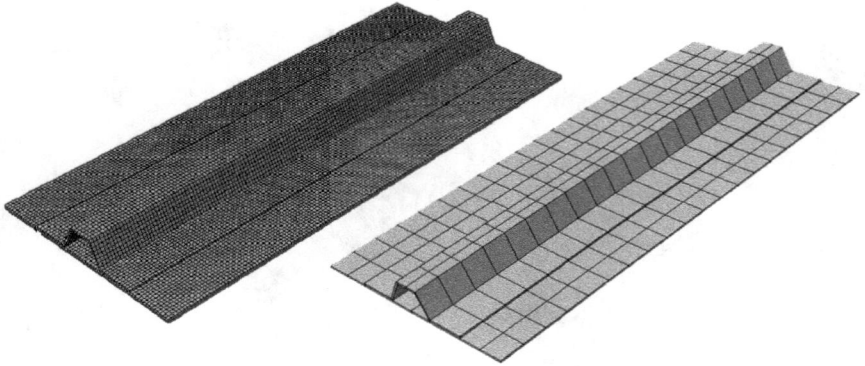

Fig. 4.3. The HFM with 4.0 mm mesh size (left) and the LFM with 30.0 mm mesh size (right).

Table 4.2. Probability distribution of random design variables.

Parameter	Probability distribution	Mean	Standard deviation
Stringer foot	Normal	43	0.215
Stringer height	Normal	30	0.15
Distance between stringer top and foot	Normal	15	0.075
Stringer top	Normal	25	0.125
Eccentricity (ϵ)	Normal	0	9.09

variables and one constraint, can be written as

$$g_{\mathrm{MF}}(X) = P_{\mathrm{cr},c} - P_{\mathrm{cr},e,\mathrm{MF}}(X1, X2, X3, X4, \epsilon) \tag{4.17}$$

where $P_{\mathrm{cr},c}$ is the minimum buckling load as a constraint, $P_{\mathrm{cr},e,\mathrm{MF}}$ is the first buckling load from the multi-fidelity model, and ϵ is the eccentricity from the centre of mass.

The structure fails when $g_{\mathrm{MF}}(X) < 0$. Each random variable has a probability distribution with its own mean and standard deviation, as seen in Table 4.2.

4.4.2.1 *Multi-fidelity modelling*

In this example, an ANN was used to create three surrogate models of the LFM and two correction response surfaces corresponding to the ratio $\beta(X)$

and the difference $\delta(X)$ between the two fidelity models. The design points to create the surrogate models using the ANN were obtained using optimal Latin hypercube sampling (OLHS). The sampling range was determined by the value of each random variable's cumulative distribution function, spanning from 0.5% to 99.5%. The reason is that the sampling points are highly concentrated in the high probability region of each random variable and less so in their low probability counterparts. Once OLHS collects the design points, the HFM and the LFM using the FEM solver calculate the corresponding output values for each design point. Then, the training datasets used to create the three different surrogate models are obtained. The training dataset comprising 11 points was sampled using OLHS because it is the minimum number of sampling points required for ANN to create the surrogate models with four design variables. In order to evaluate the quality of the surrogate models, the test dataset, comprising 30 points, was also sampled from each variable's cumulative distribution function within the same sampling range. The ANN created the following three surrogate models: the LFM, and the ratio and difference between the two fidelity models. These models were validated using the separation method considering the test dataset. The multi-fidelity models were constructed using these surrogate models and the low-fidelity FEM model, as shown in Table 4.3. The direct multi-fidelity models, MF1 and MF2 in Table 4.3, spend considerable computation time to conduct the reliability analysis because they call the first buckling load from the low-fidelity FEM models. It should be noted that the computation time of the low-fidelity FEM models is not cheap when it comes to performing thousands of FEM simulations for the reliability analysis. To improve computation time savings, the surrogate models of MF1 and MF2 were also generated using the training dataset comprising 40 points. The test dataset has 20 points from the same range of the cumulative distribution function. The indirect multi-fidelity models, MF3 and MF4, were created without calling the low-fidelity FEM models.

Table 4.3.   Multi-fidelity model.

Model	Output approximation
MF1	$Y_{\mathrm{MF1}}(X) = \beta^{\mathrm{ANN}}(X) \cdot Y_{\mathrm{LF}}^{\mathrm{FEM}}(X)$
MF2	$Y_{\mathrm{MF2}}(X) = \delta^{\mathrm{ANN}}(X) + Y_{\mathrm{LF}}^{\mathrm{FEM}}(X)$
MF3	$Y_{\mathrm{MF3}}(X) = \beta^{\mathrm{ANN}}(X) \cdot Y_{\mathrm{LF}}^{\mathrm{ANN}}(X)$
MF4	$Y_{\mathrm{MF4}}(X) = \delta^{\mathrm{ANN}}(X) + Y_{\mathrm{LF}}^{\mathrm{ANN}}(X)$

4.4.2.2 *Results and discussion*

In this section, the multi-fidelity models were used to conduct the reliability analysis of the composite panel for linear buckling under eccentric load. The reliability analysis was carried out using MCS, FORM, and SORM. A minimum buckling load constraint of 20 kN was applied, which is depicted in each figure with a dashed red line. If the buckling load of each simulation is less than this constraint, the composite panel is supposed to fail. Surrogate models using a different number of HFMs were also generated to see how accurate the multi-fidelity models are and find the equivalent number of HFMs needed to achieve a similar level of accuracy level.

The accuracy of multi-fidelity models is evaluated by comparing the mean and standard deviation at the mean value point. As shown in Tables 4.4–4.6, the reliability analysis results using the multi-fidelity models are very close to those obtained with an HF100 model. The latter, which is a surrogate model using 100 HFMs, is presumed to give the most accurate value. In addition, an HF11 model, a surrogate model that uses the minimum number of HFMs for the ANN to create the surrogate model, is

Table 4.4.   Reliability analysis results using MCS.

	HF11	HF20	HF30	HF40	HF50	HF100	HF11+LF11			
							MF1	MF2	MF3	MF4
Mean	29.57	29.2	29.56	27.99	27.97	27.88	28.58	28.58	29.41	29.55
Standard deviation	6.59	6.85	6.14	6.17	6.47	7.13	6.72	6.05	6.4	6.51
Reliability index	1.6	1.44	2.17	1.81	1.59	1.35	1.41	1.75	1.6	1.6
Error (%)	6.1	4.7	6	0.4	0.3	—	2.5	2.5	5.5	6

Table 4.5.   Reliability analysis results using FORM.

	HF11	HF20	HF30	HF40	HF50	HF100	HF11+LF11			
							MF1	MF2	MF3	MF4
Mean	29.54	29	29.44	26.47	26.36	26.48	27.31	27.46	29.3	29.52
Standard deviation	6.73	6.89	4.96	4.87	6.12	7.18	6.64	4.76	6.57	6.74
Reliability index	1.51	1.55	2.61	2.23	1.74	1.32	1.44	2.06	1.52	1.5
Error (%)	11.56	9.66	11.72	0.03	0.46	—	3.13	3.7	10.64	11.47

Table 4.6. Reliability analysis results using SORM.

| | HF11 | HF20 | HF30 | HF40 | HF50 | HF100 | HF11+LF11 | | | |
							MF1	MF2	MF3	MF4
Mean	29.54	29	29.44	26.47	26.35	26.48	27.31	27.46	29.3	29.52
Standard deviation	6.73	6.89	4.96	4.87	6.12	7.18	6.64	4.76	6.57	6.74
Reliability index	1.55	1.56	2.61	2.23	1.74	1.35	1.48	2.06	1.56	1.54
Error (%)	11.56	9.66	11.72	0.03	0.46	—	3.13	3.7	10.64	11.47

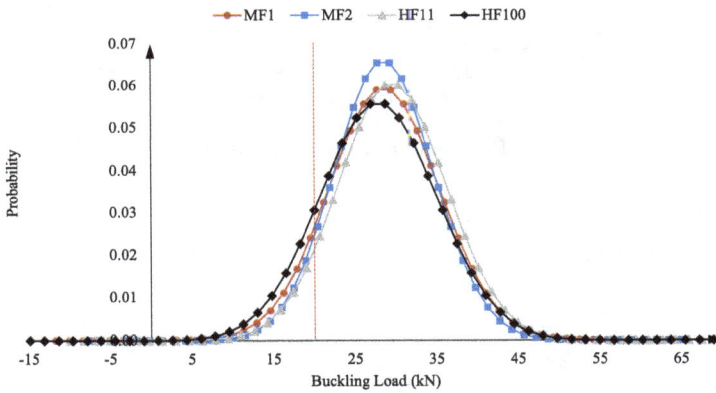

Fig. 4.4. Reliability analysis result using MCS.

tested. As the number of HFMs to generate the surrogate model decreases, the resulting surrogate models do not produce accurate solutions compared to HF100. These mean and standard deviation values are calculated from the design point results of MCS using Sobol sampling, whereas FORM and SORM calculate these two values from the output and gradient at the mean value point. Table 4.4 and Figure 4.4 show that the mean and standard deviation of HF100 are 27.88 kN and 7.13, respectively. These values are used as the most accurate values to evaluate the accuracy of the created multi-fidelity models. The difference in the mean values between HF100 and HF11 is 6.1%, while that for the direct multi-fidelity models, MF1 and MF2, is smaller at 2.5%. However, the indirect multi-fidelity models, MF3 and MF4, show similar differences to the HF11, which are 5.5% and 6.0%, respectively. The standard deviation of the HF100 using MCS is 7.13. The standard deviations of MF1 and MF2 at the mean value point are 6.72 and

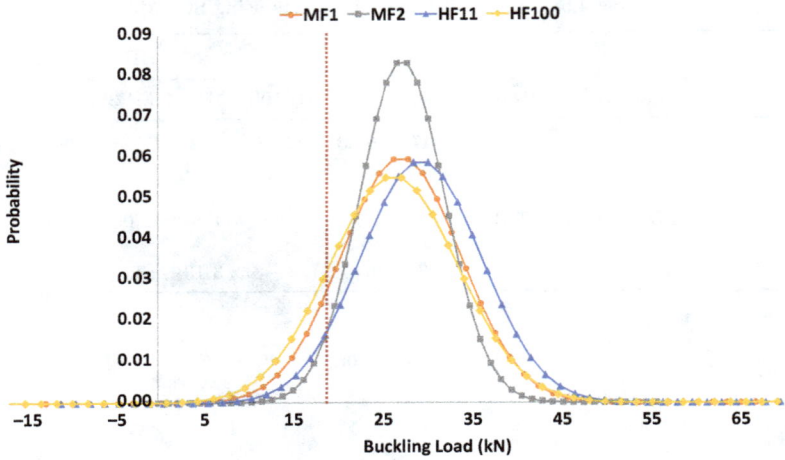

Fig. 4.5.   Reliability analysis result using FORM.

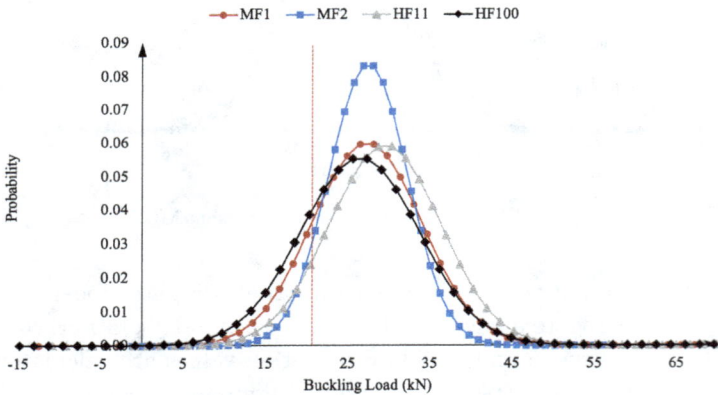

Fig. 4.6.   Reliability analysis result using SORM.

6.05, respectively, whereas MF3 and MF4 show values of 6.40 and 6.51, respectively. Figures 4.5 and 4.6 show that MF1 and MF2 show similar levels of accuracy to HF100, although the standard deviation of MF2 is smaller.

Tables 4.5 and 4.6 clearly show that the mean and standard deviation are identical for FORM and SORM since they use the same output and gradient at the mean value point. The means of FORM and SORM at the mean value point are slightly smaller than the mean of MCS, but the standard deviation is nearly the same as that of MCS. Interestingly, the different error in the mean of HF11 using FORM and SORM is over 11% compared to the mean of HF100. When the number of HFMs used to generate the surrogate model is more than 40, the differences concerning HF100 are less than 1%. MF1 and MF2 provide better accuracy than MF3 and MF4. In particular, the standard deviation of MF1 was more accurate than that of HF50. It can be seen from Tables 4.4 and 4.6 that the reliability indexes from SORM are closer to the MCS results than the FORM results. It suggests that SORM can provide a more accurate solution in the failure domain than FORM because SORM takes into account the curvature of the limit state function using second-order derivatives. In fact, it is found that the reliability index of MF1 is closer to that of HF100 than the reliability index of HF50 is to HF100. Although MF2 provides an accurate mean, its reliability index is poor because the standard deviation is smaller than those of other multi-fidelity models. MF3 and MF4 show similar levels of reliability indexes between HF11 and HF20.

In order to evaluate computation time savings, the computation time of each model was normalised by that of HF100 using MCS. The average FEM simulation times for the HFM and the LFM were 47 and 10 seconds, respectively. The average computation time for one surrogate model was 0.0057 seconds, and this computation time was calculated over 1,000,000 runs. The computation time for each reliability analysis using different models was calculated by multiplying the simulation number by the average computation time. The computation time savings are compared in Figure 4.7. MF1 and MF2 were constructed using 11 of the HFMs and 51 of the LFMs, while MF3 and MF4 were built using 11 of the HFMs and 11 of the LFMs. It is interesting to note that there were notable computation time savings in the use of multi-fidelity models. In particular, the computation times of both MF1 and MF2, which presented highly accurate solutions, are about 45% of that of HF100 using MCS. The computational cost of these two models is also cheaper than that of HF40 having an equivalent accuracy. It is seen that FORM and SORM do not show a significant difference in computation time compared to MCS because this problem converges with a small number of MCS. Suppose the problem is more complex or the reliability analysis is conducted using only high-fidelity FEM models. In that case, the computation time savings through the multi-fidelity models will increase significantly. This comparison clearly highlights that the multi-fidelity model provides not only accurate solutions that are similar to those

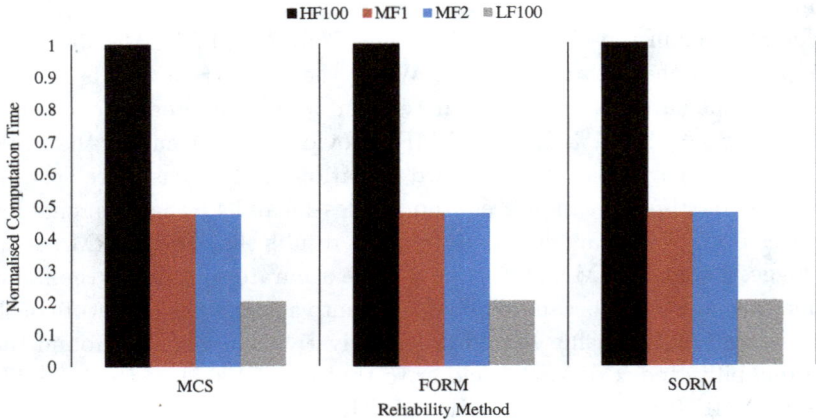

Fig. 4.7.   Computation time for different multi-fidelity models.

obtained from an HFM with many design points but also significantly lower computation times.

4.4.3 *Multi-fidelity reliability-based design optimisation*

In this section, a multi-objective probabilistic optimisation is carried out. It is demonstrated that the multi-fidelity models provide accurate solutions and high computation time savings compared to the surrogate model of the HFM. Multi-fidelity RBDO is conducted to validate the concept of multi-fidelity models in a probabilistic optimisation process, requiring a number of simulations to consider design uncertainties. The four geometric design variables of the same composite panel each have their own uncertainty, as shown in Table 4.2. NSGA-II is used, which is a multi-objective evolutionary optimisation method.

4.4.3.1 *Problem definition*

In general, RBDO includes reliability analysis in its optimisation process, and this example involves considering the uncertainties of four geometric parameters of the composite panel. The RBDO process ensures that the optimal design meets the requirements of a specific probabilistic constraint defined by a prescribed reliability index, β. In this example, random design variables are characterised by a normal distribution, and the probability failure $P_{f,\mathrm{MF}}$ is related to the prescribed reliability index β_{MF} as $P_{f,\mathrm{MF}} =$

Table 4.7.   Sobol sampling vs. simple random sampling.

	MCS Sobol sampling	MCS random sampling
Simulation number	150	600
Probability of success	1	0.9967
Mean	53.76	53.77
Standard deviation	0.28	0.28
Minimum	53.12	52.93
Maximum	54.45	54.58

$\Phi(-\beta_{\mathrm{MF}})$. As mentioned before, there are three methods to calculate the reliability of the structure: MCS, FORM, and SORM. In particular, MCS was conducted using the Sobol sampling method because simple random sampling requires a vast number of simulations during the optimisation process. According to Table 4.7, the number of simulations for MCS using the Sobol sampling is 20% fewer than those for simple random sampling. At the same time, the means and standard deviations of the statistical results are nearly identical to each other. To ensure the accuracy of FORM and SORM, the step size of the FDM was set at 0.001, and the convergence tolerance was determined to be 0.0001. The constraints of this optimisation process are the maximum mass and the target probability of failure, which are 1.0 kg and 0.00135, respectively. The objective functions are to maximise the first buckling load and to minimise the structure mass. Parameter studies using NSGA-II were carried out to set the population size and generation number; these were determined to be 12 and 60, respectively. The details of the multi-fidelity models are described in the following section.

4.4.3.2 *Multi-fidelity modelling*

The minimum number of design points for an ANN to create the surrogate models was 10 since there are four geometric design variables. The multi-fidelity models were constructed using 10 HFMs and 10 LFMs. The surrogate models using 100 HFMs, known as HF100, were also created to evaluate the performance of the multi-fidelity models. OLHS is employed to build the training and test datasets. In particular, a total of 300 design points, 100 for the training dataset and 200 for the test dataset, were collected to create the surrogate models of the HF100. A total of 30 design points — 10 for the training dataset and 20 for the test dataset — were also sampled to create two surrogate models consisting of the HFM and the LFM, respectively. These training and test datasets were also used to generate the two correction factors, $\beta(X)$ and $\delta(X)$, for the LFM to represent the response surfaces of the HFM. The design points for the training dataset were determined over the range of -20% to 20% concerning the mean of each design variable. The design points for the test dataset were selected from the broader range of -25% to 25% in order to evaluate the quality of the surrogate model. Through the training dataset from the sampling range, two direct multi-fidelity models and two indirect multi-fidelity models were constructed, as can be seen in Table 4.3. Each model has four inputs relating to the geometry and two outputs corresponding to mass and buckling load.

Table 4.8 highlights that the two direct multi-fidelity models, MF1 and MF2, showed better quality than the two indirect multi-fidelity models, MF3 and MF4. HF100 presented nearly the same response as the high-fidelity FEM models, whereas HF10, which consists of 10 HFMs, showed significant differences compared to the four multi-fidelity models. All models provided the correct mass because the response surface of mass is simple for the multi-fidelity models to represent. It is interesting to note that MF1 and MF2 provided more accurate results for buckling load and mass than HF10, even though the computation of these two models was slightly more expensive due to the extra LFMs needed to improve the quality of the multi-fidelity models. The increase in computation time caused by the LFMs is worthwhile since the multi-fidelity models provide more accurate results than HF10. If the number of HFMs increases until their accuracy

Table 4.8.   Multi-fidelity model validation.

| Model | Fitness error | |
	Buckling load	Mass
MF1	0.0101	0.0009
MF2	0.0107	0.0009
MF3	0.0155	0.0034
MF4	0.0155	0.0034
HF10	0.0159	0.0029
HF100	0.0034	0.0006

matches that of the multi-fidelity models, the computation time resulting from this increase should be much higher than that of the multi-fidelity models. It is seen that the errors of the indirect multi-fidelity models were higher than those of the direct multi-fidelity models because of their reliance on surrogate models based on the design points from the LFM. These four multi-fidelity models were validated to conduct the RBDO process as an alternative to using the HFM.

4.4.3.3 *Results and discussion*

In Figures 4.8 and 4.9, the optimisation results of the multi-fidelity models using FORM are compared with those of the high-fidelity modelling-based surrogate model (HF100). The results using MCS and SORM are not presented in this example because they are nearly identical to those obtained using FORM. As shown in these figures, the Pareto fronts show the optimal design results that satisfy the desired objectives and constraints. It should be noted that the slope of the Pareto front line is changed when the structure mass is around 0.94 kg. It means that the first buckling load increases gradually until the mass reaches 0.94 kg. However, when the mass exceeds 0.94 kg, the increase in buckling load is not as prounounced as the increase in structure mass. It is determined that the design geometries for the mass of 0.94 kg are reasonable design values in the given design space.

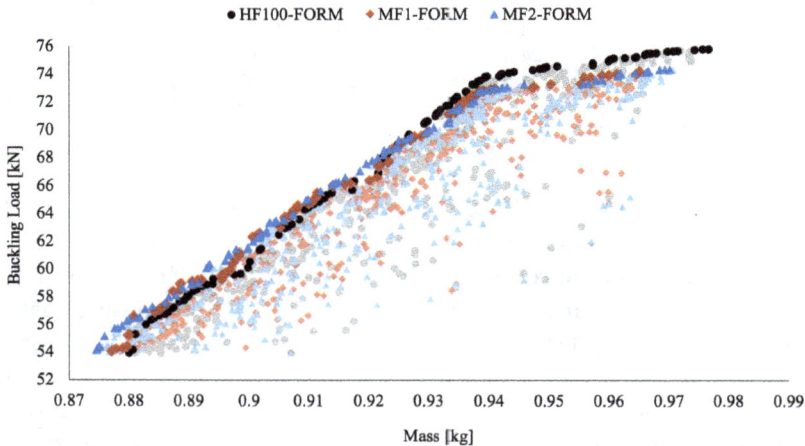

Fig. 4.8. Comparison to RBDO results using FORM (HF100 vs. direct multi-fidelity models).

Note: Bold points are the Pareto fronts while fainter points are feasible solutions.

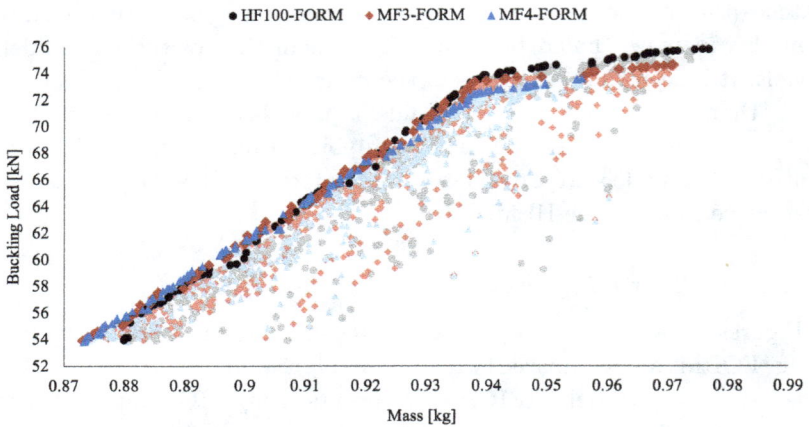

Fig. 4.9.   Comparison to RBDO results using FORM (HF100 vs. indirect multi-fidelity models).

Note: Bold points are the Pareto fronts while fainter points are feasible solutions.

Table 4.9.   Initial and chosen geometry of the composite panel.

Model	X1 (mm)	X2 (mm)	X3 (mm)	X4 (mm)	Mean (kN)	STD	Mass (kg)
Initial	43	30	15	25	53.76	—	0.9
HF100	51.6	24.3	18	30	73.86	0.4	0.94
MF1	51.6	24.5	18	30	72.87	0.36	0.94
MF2	51.5	24.4	18	30	72.83	0.36	0.94
MF3	51.6	24.5	18	30	73.45	0.39	0.94
MF4	51.6	24.6	18	30	72.67	0.36	0.94

Table 4.9 and Figure 4.10 compare the chosen geometries, showing that when the mass of the composite panel is 0.94 kg, the linear buckling load is at its economically maximum value. The results of HF100 are the most accurate because this model consists of sufficient HFMs to ensure that the correct first buckling load value is obtained. It is worth noting that the chosen optimal geometry values from the multi-fidelity models are nearly identical to those from HF100. The mean and standard deviation of the multi-fidelity models are similar to those of HF100 for the same mass. Figures 4.10 and 4.11 show the probabilistic distribution of each multi-fidelity model and the optimal geometric design from RBDO. The direct

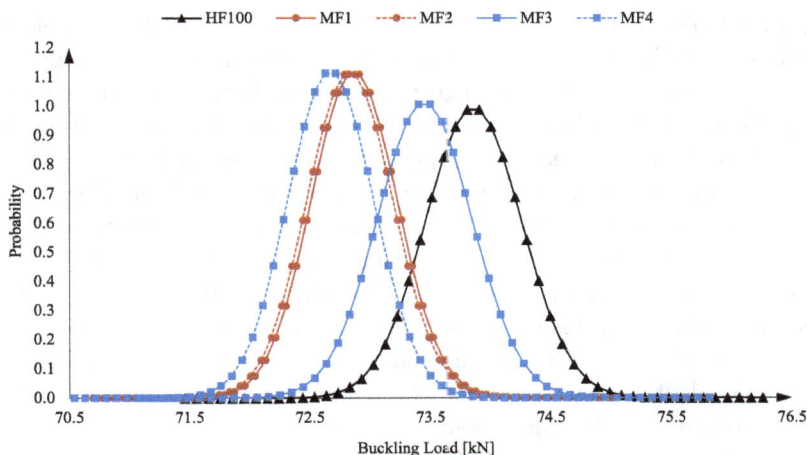

Fig. 4.10. Reliability-based design optimisation results.

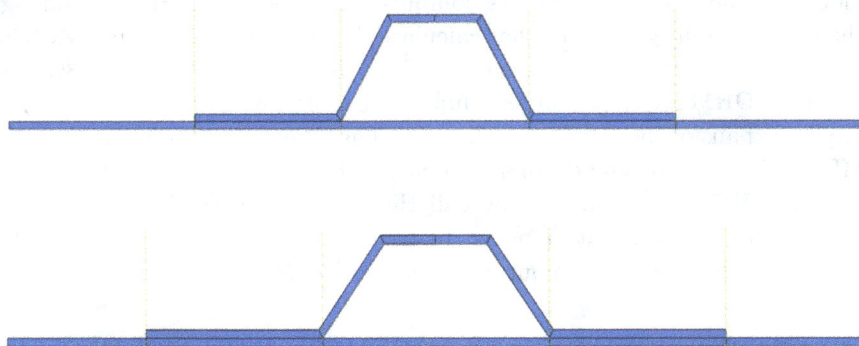

Fig. 4.11. Mono-stringer stiffened panel geometry optimised for maximum linear buckling load based on 0.94 kg: (top) Initial model and (bottom) RBDO model (MF1).

multi-fidelity models, MF1 and MF2, have almost the same mean and standard deviation. The probabilistic distributions of indirect multi-fidelity models, MF3 and MF4, show a slightly different mean of the first buckling loads, although they have nearly the same standard deviation. Therefore, the accuracy of all multi-fidelity models was validated.

The main goals of this study are to achieve computation time savings and maintain accuracy. It is essential to show how much computation time savings can be achieved by using multi-fidelity models. In order to enable a reasonable comparison of the computation time between the models, the required number of simulations for each model during this optimisation process is compared. The average computation time of HFM and LFM using Abaqus/CAE was calculated over 100 runs for each. The computation times were 47 and 10 seconds, respectively. The computation time of one surrogate model was assumed to be given by the total simulation time divided by the total number of simulations used in the whole optimisation process. The total simulation time using an Intel Core i7-6700 CPU @ 3.40 GHz and the number of simulations using all surrogate models were 7,036 seconds and 1,229,085, respectively. The computation time of one surrogate model was 5.7 ms (millisecond). For each model, the computation time was calculated by combining the number of simulations required for the surrogate models, high-fidelity FEM models, and low-fidelity FEM models. As shown in Figure 4.12, all computation times were normalised by the computation time of the HF100 using MCS, which is the most computationally expensive. This figure clearly shows that the multi-fidelity models require considerably less computational cost than HF100. Among the three reliability methods that calculate the probability of failure, MCS is the most computationally expensive. FORM is a little cheaper than SORM because SORM requires a larger number of simulations for the second-order Taylor expansion in the failure domain. The direct multi-fidelity models, MF1 and MF2, are slightly more expensive than the indirect multi-fidelity models, MF3 and MF4, as they call the low-fidelity FEM models when they create the surrogate models. In particular, all multi-fidelity models show a similar level of computation time to LF100, which consists of 100

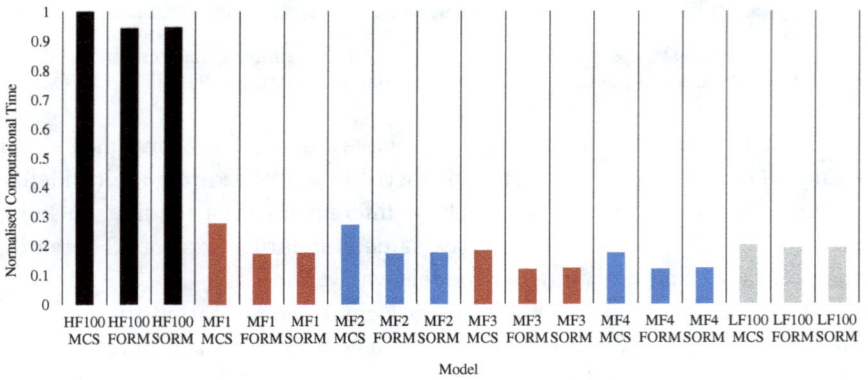

Fig. 4.12.   Computation time for different multi-fidelity models.

low-fidelity FEM models. It should be noted that the computation time of the multi-fidelity models is reduced by at least 70% compared to that of HF100. If the optimisation is conducted using the high-fidelity FEM models without the surrogate models, the multi-fidelity models will save computation time by considerably more than 70%.

4.5 Summary

In conclusion, a multi-fidelity formulation has been developed for reliability analysis and RBDO of composite structures to consider the influence of uncertainties in design variables. This formulation enables multi-fidelity models to offer remarkable accuracy, nearly identical to that of HFMs, as well as significant computational time savings, similar to LFMs. This is accomplished by creating the response surfaces from the ratio and difference between two different fidelity models. These two models, covering the same design spaces, require the same number of design points to train the surrogate model using an ANN. The direct and indirect multi-fidelity models, depending on how the low-fidelity FEM models are used during the modelling process, are constructed. Then, these models are employed in the proposed multi-fidelity RBDO process.

Two numerical examples demonstrated the performance of the multi-fidelity models: reliability analysis and RBDO. The multi-fidelity reliability analyses considering design uncertainties were conducted using MCS, FORM, and SORM, which calculated the reliability of the structure according to the given limit state function. In the example of multi-fidelity reliability analysis, the direct multi-fidelity models, particularly the one using the ratio response surface, provided a highly accurate solution. The computational cost of the multi-fidelity models is equivalent to that of using 40 high-fidelity FEM models. These multi-fidelity models furnished a computation time saving of over 50% compared to a conventional computationally expensive method that used only an HFM. The concept of multi-fidelity modelling was also applied to the probabilistic multi-objective optimisation problem. The direct and indirect multi-fidelity models provided very close optimisation solutions to those obtained by the conventional method. The computation time using MCS was decreased by at least 70%. More substantial time savings were achieved when FORM and SORM were used in the optimisation process. These results suggest that the new multi-fidelity framework can be applied to the reliability analysis and RBDO of composite structures exhibiting design uncertainties. This framework provides a high level of accuracy and considerable computational time savings when compared to the conventional method that relies solely on HFMs.

Chapter 5

Multi-Fidelity Robust Design Optimisation Using Successive High-Fidelity Correction

It is widely acknowledged that composite structures attract attention in different industrial areas because they provide remarkable strength, stiffness, and energy savings, consistent with both environmentally friendly designs and cost-effective operations. The manufacturing process of composite structures is complex and carries various uncertainties, which influence the quality and performance of the final design. The variation in design parameters, in general, provokes unexpected deviations across the entire lifecycle, including design, manufacturing, service, and ageing. Robust design optimisation (RDO), a type of probabilistic design optimisation, is an essential design approach since it considers how the uncertainties associated with design and manufacturing could affect product quality and performance. The main idea of RDO was proposed by Taguchi Genichi [3], and it aims to deliver more stable output performance regarding the variations in design parameters, such as geometry, service environment, and material properties. RDO minimises the variation in product performance caused by design uncertainties, preventing the final design from being overly conservative due to excessive safety factors or worst-case scenarios. Hence, RDO improves the quality of the design product by stabilising the deviations in response behaviour without removing their design sources and minimising the effects of design uncertainties [53].

RDO has been studied and applied to various engineering design fields, such as structural design using the finite element method (FEM) [4, 28, 81–83] and aerodynamic design using computational fluid dynamics (CFD) [76, 84–86]. In general, RDO requires computationally expensive efforts to find robust design solutions because a large number of computational simulations must be conducted to calculate the effects of design uncertainties. This is a typical challenge to be addressed in the area of probabilistic design optimisation. Similar to RBDO in the previous chapter, the use of the surrogate model has been actively developed to tackle the computational challenge posed by the RDO process [28]. Most of them approximate the expensive computation models using artificial neural networks (ANNs), kriging, polynomial regression, or the response surface method (RSM) [4, 76, 87–89]. One notable example concerning the RDO of composite structures using surrogate models is the RDO framework for a composite stiffened panel under nonlinear progressive failure analysis, proposed using an ANN [4]. The quality of the created surrogate model was evaluated using the cross-validation error method to determine whether it provides computation time savings while maintaining an acceptable level of accuracy. This model was applied to conduct a robustness analysis considering the design uncertainties of the composite stiffened panel. This work allows the concept of RDO to be implemented in the design of composite structures for nonlinear damage progressive problems. It shows the potential to broaden the application area of RDO using a surrogate model.

As discussed in Chapter 4, there are different types of multi-fidelity modelling methods that have been developed to improve the computational efficiency of conventional high-fidelity surrogate modelling methods. When a problem becomes large-scale and complex, the computational cost to create the surrogate model is still extremely expensive because a large number of high-fidelity models (HFMs) must be simulated using a numerical solver. However, the computational cost for running even a single high-fidelity FEM simulation of the composite structure is too high to neglect when the problem exhibits nonlinear response behaviour or involves many design variables. Multi-fidelity modelling methods developed so far have been used to conduct optimisation and uncertainty propagation using the combination of both HFMs and low-fidelity models (LFMs). In particular, the vast majority of approaches combining the two different fidelity models use response correction methods for the LFM to represent the response behaviour of the HFM [13, 29, 56]. Different multi-fidelity modelling methods, such as the co-kriging method, have been applied to aerodynamic optimisation problems [72, 84, 90]. Even though these multi-fidelity modelling methods provide acceptable accuracy and computational efficiency compared to conventional high-fidelity surrogate modelling methods, they have only been demonstrated in mathematical examples, aerodynamic design optimisation, and structural design optimisation of isotropic materials. The primary limitation of using these methods is the difficulty in handling large-scale problems having many design variables. In order to create the response correction functions, such as the ratio or difference between the HFM and the LFM, many high-fidelity simulations should be conducted using FEM or CFD solvers to build a correct response surface. This causes a tremendous computational burden when the system has many design variables. It should be noted that the number of high-fidelity simulations must be equal to that of low-fidelity simulations to construct a multi-fidelity model using the response correction methods.

The main objective of this work is to develop a novel multi-level multi-fidelity modelling-based RDO framework. It is then applied to a probabilistic optimisation problem when the HFM and the LFM possess an unequal number of design variables so that the computational cost to construct the multi-fidelity model is reduced. This developed multi-fidelity RDO framework requires a smaller number of high-fidelity FEM simulations compared to conventional multi-fidelity modelling methods that require the same number of FEM simulations for the HFM and LFM. To achieve this research objective, the HFM has fewer design variables during the optimisation process, while the LFM has more design variables to explore the whole design space while sharing the same design variables with the HFM. Multi-level optimisation was also considered to achieve greater computational efficiency by dividing the optimisation problems into several subproblems. This framework is driven by the LFM covering the whole design space while updating the optimal solutions using high-fidelity corrections. Its applications demonstrated this approach to both deterministic optimisation (DO) and RDO of a mono-stringer stiffened composite panel under the post-buckling regime. Finally, the robust and deterministic designs were compared by conventional optimisation methods using the HFM. The performance of this new framework was evaluated in terms of optimisation solution accuracy and computation time savings.

This chapter first explains the concept of the multi-fidelity modelling method, which consists of different design spaces between the HFM and the LFM. Then, it introduces the developed multi-fidelity formulation and describes how this formulation implements the probabilistic design optimisation. At the end of this chapter, two numerical examples of DO and RDO are illustrated to design a composite structure for demonstrating the new multi-fidelity optimisation framework. The work presented in this chapter is based on the authors' work [51].

5.1 Multi-Fidelity Models in Different Design Spaces

As discussed in Chapter 3, the traditional multi-fidelity methods developed so far can be categorised based on how they allow the LFM to represent the response behaviours. The majority of these multi-fidelity methods set up the response correction function, such as multiplicative, additive, and comprehensive correction methods. These methods require the same number of high-fidelity and low-fidelity FEM simulations. Some methods using Gaussian processes (GPs), including co-kriging, require fewer high-fidelity FEM simulations than low-fidelity FEM simulations. However, the design variables in the HFM and LFM must be identical to each other. These traditional methods may cause significant computational challenges when the optimisation problem is of large scale or has many design variables. The space mapping method, which creates a proper transformation function between different design spaces, has not been used in the field of structural optimisation so far due to its limitations. The limitations will be described in the following section.

5.1.1 *Space mapping method in structural optimisation*

The space mapping method allows different fidelity models to cover different design spaces, and it constructs the multi-fidelity model. This method builds a mapping function between the HFM comprising all design variables and the LFM comprising a few design variables. The mapping function enables the optimisation results obtained in the low-fidelity design space to be transformed into the high-fidelity design space using the inverse mapping function. The idea of the space mapping method was first proposed to design the microwave circuit [71]. Then, its application scope has expanded to the field of aerodynamic design [91]. However, this method has been scarcely applied in the area of structural optimisation. The only example used the space mapping method to conduct a DO of a steel beam structure with two design variables subjected to a uniformly distributed load [92]. However, the space mapping method considered in the vast majority of related literature shared the same number of design variables between the HFM and the LFM [93]. Even though the HFM and LFM have different numbers of design variables, the fidelity is not the level of discretisation but rather that of numerical solvers [91].

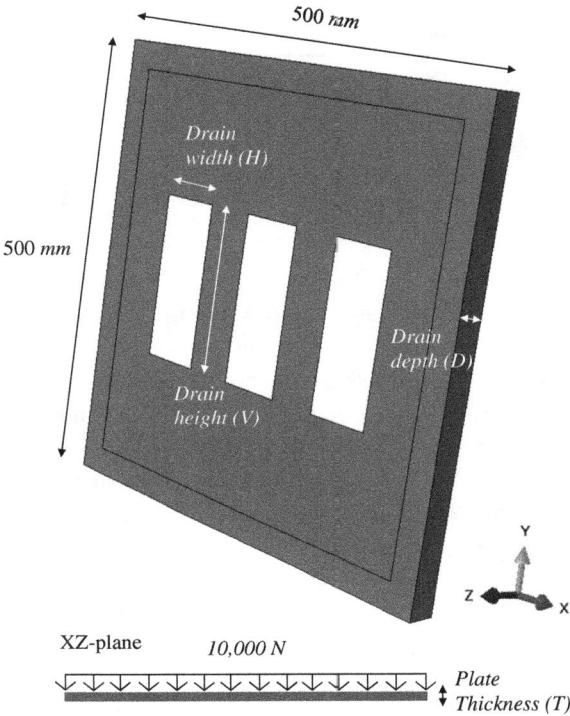

Fig. 5.1. Isotropic steel drain cover with four design variables.

This section demonstrates whether the space mapping method is available to construct a proper multi-fidelity model for structural optimisation when the HFM and LFM possess different numbers of design variables. The numerical example is the same as the isotropic steel plate presented in Chapter 3; however, it has more design variables, as shown in Figure 5.1. The drain cover has four design variables: horizontal drain, vertical drain, drain depth, and plate thickness. This DO problem aims to minimise the mass and vertical displacement under a uniformly distributed load. The constraint is the maximum allowable stress of the drain cover. The finite element method (FEM) model of the drain cover was created using Abaqus/CAE [66]. The design ranges of each variable are described in Table 5.1.

In this example, the HFM has four design variables, H, V, D, T, while the LFM has two, v and t. The two variables for the LFM were determined to be the more dominant input design variables in relation to the objective functions compared to the other design variables. Both the high- and

Table 5.1.   Design range of each variable.

Parameter	Value (mm)
Drain height	$144 < V < 216$
Drain width	$56 < H < 84$
Drain depth	$24 < D < 36$
Plate thickness	$4 < T < 6$

low-fidelity FEM models were created using the identical mesh size, although they carry different design variables.

As illustrated in Figure 5.2, the optimisation process using the space mapping method consists of two steps: a mapping parameter extraction and the DO process. Basically, X_{HFM} is a vector of high-fidelity design variables, and the mapping function, P, transforms the X_{HFM} into X_{LFM}, which is a vector of low-fidelity design variables. Then, the optimisation problem can find optimal solutions in the low-fidelity design spaces. Finally, the low-fidelity optimal solutions, X^*_{LFM}, are transformed into high-fidelity optimal solutions, X^*_{HFM}, using the inverse mapping function. The parameter extraction is the main step for the space mapping method to build a multi-fidelity model.

In this example, the created mapping function transforms the high-fidelity design spaces having four design variables into the low-fidelity design spaces having two design variables. First, n high-fidelity training points should be collected from the entire design space using design of experiments (DoE). Then, the output responses are obtained to set up the high-fidelity training dataset. A linear mapping function, P, should be presumed, which can be represented as $X_{\text{LFM}} = P(X_{\text{HFM}}) = B \times X_{\text{HFM}} + C$. It should be noted that the dimensions of the mapping matrix B and the vector C are (2×4) and (2×1), respectively, relying on a different number of design variables between the HFM and the LFM. Parameter extraction is a sub-optimisation problem to obtain each mapping parameter of the matrix B and the vector C. This sub-optimisation problem minimises the difference between the high-fidelity response, $Y^*_{\text{HFM}}(X^*_{\text{HFM}})$, and the low-fidelity response, $Y^*_{\text{LFM}}(X^*_{\text{LFM}})=Y_{\text{LFM}}(B \times X_{\text{HFM}} + C)$. The sub-optimisation offers the mapping parameters. Then, the surrogate model, $\hat{P}$, of the mapping parameters can be established using an ANN. This surrogate model satisfies with $Y_{\text{LFM}}(\hat{P}(X_{\text{HFM}})) \approx \hat{Y}_{\text{HFM}}(X_{\text{HFM}})$ and is used to carry out the optimisation problem until the low-fidelity optimal solutions, Y^*_{LFM},

Space mapping parameter extraction

Sample high-fidelity design points (i.e., OLHS) to set up input and output datasets X_{HFM} and Y_{HFM}

Create the high-fidelity surrogate model using ANNs $Y_{HFM}(X_{HFM})$

Sub-optimisation to find mapping parameters in the matrix B and vector C

Find B and C
$$X_{LFM} = P(X_{HFM}) = B \times X_{HFM} + C$$

Minimise
$$Y_{HFM}(X_{HFM}) - Y_{LFM}(B \times X_{HFM} + C)$$

Create a surrogate model P of mapping function using the obtained parameters

High-fidelity design variables X_{HFM}

Space mapping function

Low-fidelity design variables X_{LFM}

Deterministic optimisation process using LFM and mapping function P

Find the optimal solutions X^*_{LFM} at the low-fidelity design space

Transform them to high-fidelity solutions X^*_{HFM}

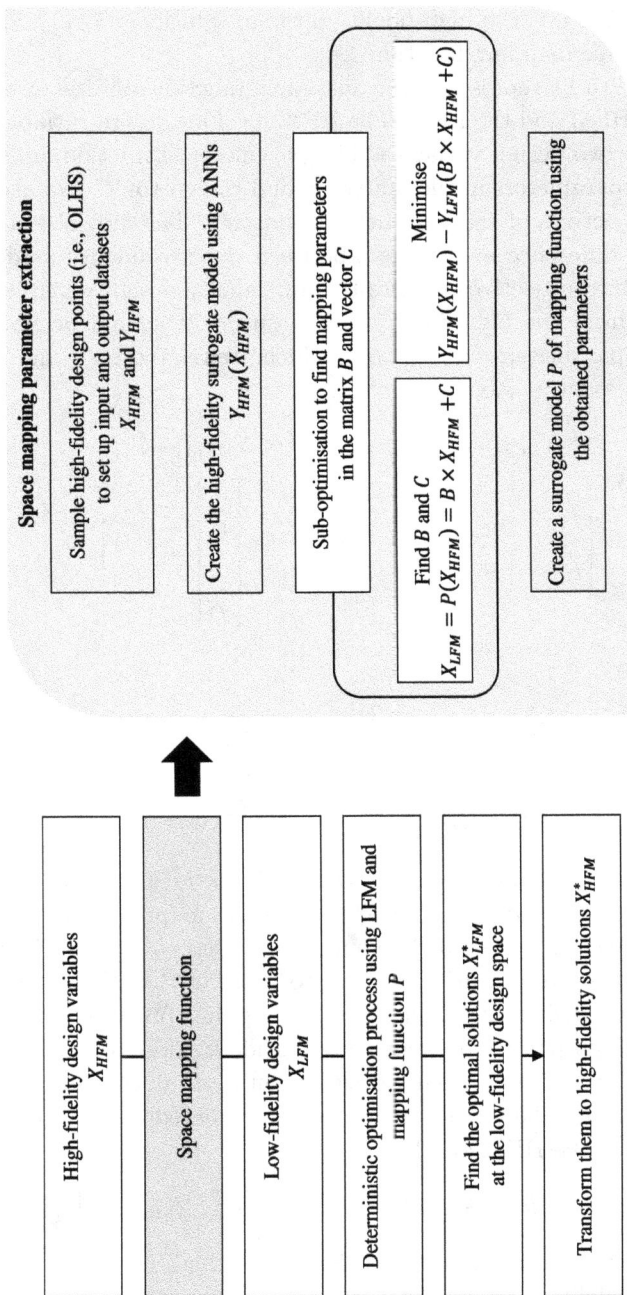

Fig. 5.2.   Deterministic optimisation process using the space mapping method.

are obtained. Finally, the high-fidelity optimal solutions, Y_{HFM}^*, are found using the inverse mapping function.

Equation (5.1) represents the mapping function of design variables between the HFM and the LFM. The HFM has four design variables, while the LFM has two design variables. The parameter extraction aims to find the mapping parameters in the matrix B and the vector C. Equation (5.2) shows the objectives of the parameter extraction. The first objective is to minimise the difference in responses between the two fidelity models. The second and third objectives require the low-fidelity design variables to not be far away from the high-fidelity design space. It should be noted that the mapping parameters B_{12} and B_{23}, which are related to V and T in the HFM, are constant at 1.0:

$$X_{\mathrm{LFM}} = P(X_{\mathrm{HFM}}) = B \times X_{\mathrm{HFM}} + C \tag{5.1}$$

$$\begin{bmatrix} v \\ t \end{bmatrix} = \begin{bmatrix} B_{11} & B_{12} & B_{13} & B_{14} \\ B_{21} & B_{22} & B_{23} & B_{24} \end{bmatrix} \begin{bmatrix} H \\ V \\ T \\ D \end{bmatrix} + \begin{bmatrix} C_1 \\ C_2 \end{bmatrix}$$

Minimise:

$$\|Y_{\mathrm{HFM}}(X_{\mathrm{HFM}}) - Y_{\mathrm{LFM}}(B \times X_{\mathrm{HFM}} + C) \tag{5.2}$$

$$B_{11} \cdot H + B_{13} \cdot T + B_{14} \cdot D = 0$$

$$B_{21} \cdot H + B_{22} \cdot V + B_{24} \cdot D = 0$$

When the surrogate model of the HFM is created, the parameter extraction is conducted using NSGA-II and the mapping parameters are found, as shown in Figure 5.3. The output responses of the LFM using the mapping function were nearly identical to those of the HFM. However, it should be evaluated whether the inverse mapping function transforms a low-fidelity design point into a correct high-fidelity design point. Equation (5.3) denotes how the high-fidelity design point is obtained using the inverse mapping function. The pseudo-inverse was considered due to the different design spaces of the HFM and LFM:

$$X_{\mathrm{HFM}} = P^{-1}(X_{\mathrm{LFM}}) = \left[(B^T B)^{-1} B^T \right] [X_{\mathrm{LFM}} - C] \tag{5.3}$$

In the first case in Figure 5.3, the inverse high-fidelity design point from the low-fidelity design point was not even close to the original design

High-fidelity design space

D	H	V	T
30.78	71.33	174.73	5.23
29.22	71.09	183.42	4.77
29.84	70.85	173.48	4.78
30.57	68.67	171	5.04
30.05	70.12	179.69	4.99

Space mapping

B11	B12	B13	B14	B21	B22	B23	B24	C1	C2
0.09	1.00	0.39	-0.48	0.32	-0.17	1.00	0.23	6.76	0.08
0.09	1.00	0.25	-0.33	-0.5	0.16	1.00	0.2	1.84	0.03
0.04	1.00	0.16	-0.28	0.36	-0.14	1.00	-0.05	3.56	0.25
-0.08	1.00	0.29	-0.07	0.28	-0.06	1.00	-0.32	5.76	0.06
0.15	1.00	-0.19	-0.35	-0.32	0.09	1.00	0.23	0.61	0.13

Low-fidelity design space

v	t
174.94	5.06
183.35	4.71
172.46	4.39
170.75	4.54
179.11	5.13

Output from the high-fidelity FEM

Mass	Maximum displacement	Maximum stress
0.01	3.60	134.93
0.01	3.66	136.36
0.01	3.60	135.08
0.01	3.59	134.63
0.01	3.64	135.73

Minimised error

$$Y_{LFM}\left(P(X_{HFM})\right) \cong Y_{HFM}(X_{HFM})$$

Output from the low-fidelity FEM

Mass	Maximum displacement	Maximum stress
0.01	3.59	134.93
0.01	3.68	136.36
0.01	3.60	135.08
0.01	3.58	134.63
0.01	3.64	135.72

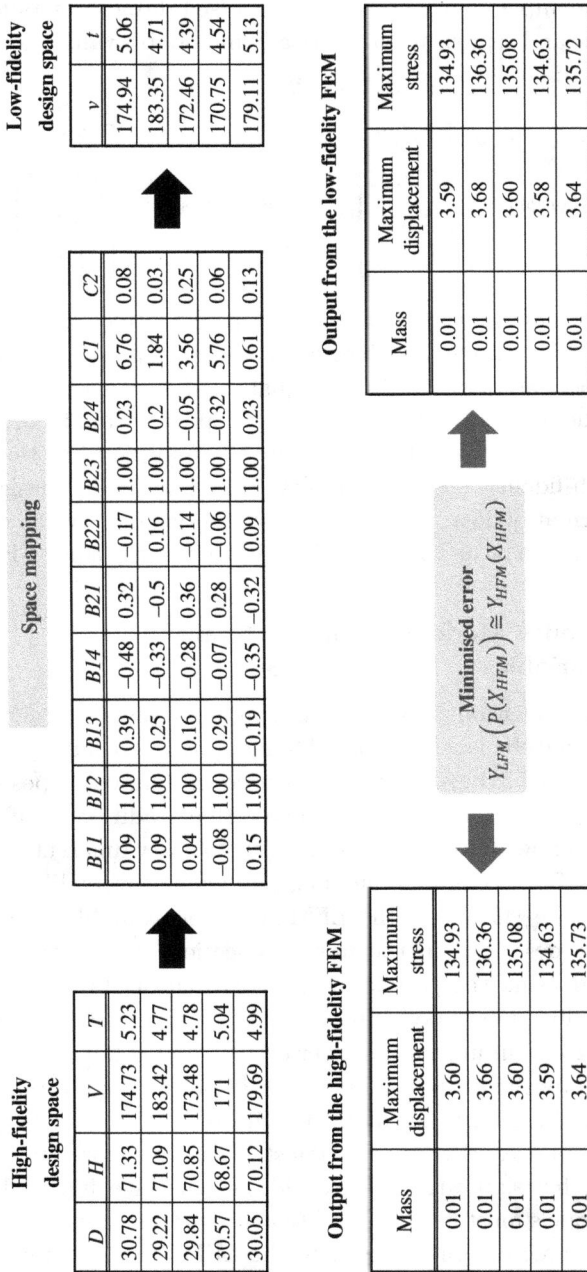

Fig. 5.3.   Space mapping method between the HFM and the LFM.

point but was completely different. The inversed design points are out of the design space and represent geometrically impossible values, as described in equation (5.4):

$$X_{\text{HFM,original}} = \begin{bmatrix} 71.33 \\ 174.73 \\ 5.23 \\ 30.78 \end{bmatrix}, \quad X_{\text{HFM,inversed}} = \begin{bmatrix} 7.89 \\ 123.72 \\ 37.93 \\ -60.82 \end{bmatrix} \quad (5.4)$$

This result shows that the inverse mapping matrix does not provide correct high-fidelity design variables when the design spaces of the HFM and LFM differ. The matrix $B^T B$ in equation (5.3) is close to a singular matrix, and the matrix $(B^T B)^{-1} B^T$ might provide an incorrect high-fidelity design point. It is not possible for the mapping function to offer reasonable high-fidelity design variables from low-fidelity design spaces because of information loss between the two fidelity models when the design variables are dependent on the mesh size when using the same FEM solver.

5.2 Multi-Fidelity Modelling with Different Design Variables

The multi-fidelity methods developed in computer science provide significant computational efficiency compared to metamodels or surrogate models. Such surrogate models have been widely used in different types of design problems in engineering. However, it is found that traditional multi-fidelity methods do not allow different fidelity models to carry different numbers of design variables. The space mapping method, which covers different design spaces between the HFM and the LFM, was not available for structural optimisation, as described in the previous section. The inverse mapping function derived using the pseudo-inverse formulation does not provide a correct high-fidelity design point from a low-fidelity design point since the mapping matrix has an unequal number of rows and columns. It should be noted that a composite structure used in large-scale design problems, such as aircraft or wind turbines, has many design variables, including geometric and material properties. For instance, if a structure has more than 30 design variables and each design variable has a wide design range, hundreds of high-fidelity FEM models must be simulated to construct the response correction function in the same design spaces with the LFM. This requires the same number of design points as the LFM to cover the entire design space, leading to considerable computational costs. Although traditional multi-fidelity methods offer a certain level of computation time savings, it should be highlighted that changes in the mesh regarding different geometry parameters in the HFM cause additional computational costs when it comes to large-scale structures.

5.2.1 *Multi-level optimisation for multi-fidelity modelling*

As engineering systems in the field of structural optimisation become more complex and enormous, this results in an increase in the number of design variables that need to be considered. As mentioned in the previous section, the computational cost of considering all design variables in high-fidelity FEM simulations is prohibitive to construct even multi-fidelity models.

In order to resolve this challenge, the developed multi-fidelity formulation adapts the concept of multi-level optimisation to improve computational efficiency compared with traditional surrogate methods and other multi-fidelity methods. Multi-level optimisation presents a wide range of benefits when it comes to large-scale optimisation problems. This optimisation approach separates a given large-scale optimisation problem into several scaled-down problems [94]. This approach breaks down the large-scale optimisation problem into several levels of design adjustments, corresponding to various collections of design variables. This adaptation ensures that the multi-fidelity formulation can offer significant computational time savings while safeguarding the probabilistic optimisation process against information loss at various levels [51].

The selection of levels and the associated design variables depends on the problem characteristics. For instance, if the level is determined using numerical analysis methods, the design variables should be defined according to the requirements of the numerical solver [95]. In comparison, if the level is defined by different objective functions, each level should include different design variables which deliver critical impact [96]. The level can also be determined by different design approaches, such as preliminary and detailed designs [4]. These levels provide many benefits when handling large-scale problems in terms of computational efficiency. It is not surprising that each level has its own objectives and constraints, depending on the design variables. In this multi-level optimisation, the designer should decide on finding reasonable solutions at each level that maximise the performance of the whole system. A typical multi-level multi-objective optimisation process can be expressed as in equation (5.5):

Level 1:

$$\text{minimise/maximise} \quad f_i(X) \quad (i = 1, 2, \ldots, I)$$

$$\text{subject to} \quad g_j(X) \leq 0 \quad (j = 1, 2, \ldots, J)$$

$$h_k(X) = 0 \quad (k = 1, 2, \ldots, K)$$

$$x_l^{(L)} \leq x_l \leq x_l^{(U)} \quad (l = 1, 2, \ldots, L)$$

Level n: (5.5)

$$\text{minimise/maximise} \quad f_0(Y) \quad (o = 1, 2, \ldots, O)$$

$$\text{subject to} \quad g_p(Y) \leq 0 \quad (p = 1, 2, \ldots, P)$$

$$h_q(Y) = 0 \quad (q = 1, 2, \ldots, Q)$$

$$y_r^{(L)} \leq y_r \leq y_r^{(U)} \quad (r = 1, 2, \ldots, R)$$

Here, X and Y are the design variables at Level 1 and Level n, respectively. f_i and f_0 are the objective functions at each level; g_j and g_p, and h_k and h_q are the inequality and equality constraints at each level; $x_l^{(L)}$ and $x_l^{(U)}$, and $y_r^{(L)}$ and $y_r^{(U)}$ are the bounds on the design variables at each respective level.

5.2.2 *Multi-fidelity formulation using high-fidelity correction*

This section introduces a new multi-level multi-fidelity modelling method to address the computational challenge and offer more computational efficiency than traditional multi-fidelity methods. This multi-fidelity method aims to manage different design spaces between the HFM and the LFM. The multi-level optimisation approach is incorporated into this developed modelling process. This method employs an ANN, which primarily incorporates the radial basis functions introduced in Section 3.1.2.1, to create the surrogate models.

Figure 5.4 represents the main idea of this multi-fidelity modelling method, which is to use the LFM with successive high-fidelity corrections across the optimisation level. Here, the HFM has fewer design variables (one variable in this example) than the LFM at each level, reducing the

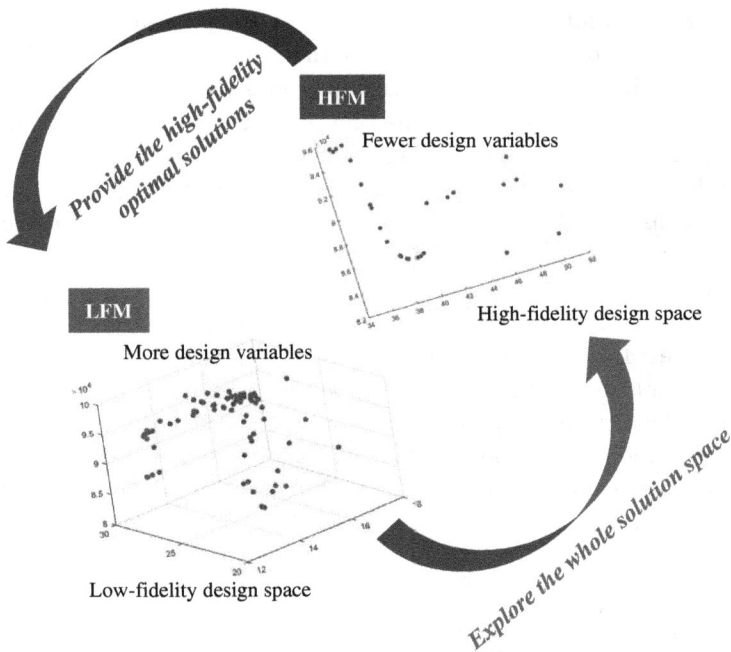

Fig. 5.4.   Concept of multi-fidelity modelling method with a different number of design variables.

number of high-fidelity FEM simulations needed to construct surrogate models. In contrast, the LFM has more design variables than the HFM at the same level. The multi-fidelity model constructed by these two fidelity models encompasses different dimensions of design spaces, and it is utilised at each level of the probabilistic optimisation process. Fundamentally, the optimal solutions of high-fidelity design variables are obtained using the optimisation loop of the HFM. At the same time, the solution spaces of other design variables that are not included in the HFM can be explored by the optimisation loop of the LFM having all design variables. After every optimisation level, both the HFM and the LFM are corrected using the optimal solutions from the multi-fidelity model at each level, and then the updated models move on to the following optimisation level. These updated models are exploited to construct an improved multi-fidelity model that carries new high-fidelity design variables during the following optimisation process. The optimal solutions of the low-fidelity design variables at the previous level are employed by the initial starting points of the next level to discover global solutions efficiently. The LFM can complement information loss in the whole design space caused by the HFM not embracing the entire design space. In this manner, the HFM provides accurate optimal solutions of high-fidelity design variables. The LFM explores the whole solution space of all design variables while sharing them with the HFM.

As defined in Table 5.2, the formulation highlights how a multi-fidelity model is constructed using a different number of design variables between the HFM and the LFM. At the first level, m design variables in the HFM are selected from all design variables, n. These m selected design variables are defined by $H^{(I)}$. Other design variables not chosen are fixed at their initial values. In comparison, the LFM has n design variables defined by $L^{(I)}$. It is not surprising that $H^{(I)}$ is a subset of $L^{(I)}$. This enables the constructed multi-fidelity model to explore the whole design space using the LFM. After the optimisation at this level, the optimal solutions are found, which are represented by $\hat{H}^{(I)}$ and $\hat{L}^{(I)}$ of high-fidelity design variables and low-fidelity design variables, respectively. Then, the selected high-fidelity

Table 5.2.   Formulations of multi-level multi-fidelity modelling method.

	HFM at Level I	LFM at Level I
Design variables	$H^{(I)}=[x^{(1)},...,x^{(m)}]$, $(m<n)$ $x^{(m+1)},\ldots,x^{(n)}$ are fixed	$L^{(I)} = X = [x^{(1)}, x^{(2)},\ldots, x^{(n)}]$ $H^{(I)} \subseteq L^{(I)}$
Number of design variables	m	n (all design variables)
Optimal values	$\hat{H}^{(I)} = [\hat{x}^{(1)},\ldots,\hat{x}^{(m)}]$	$\hat{L}^{(I)} = [\hat{x}^{(1)},\ldots,\hat{x}^{(n)}]$
Corrected HFM & LFM	$X = [\hat{H}^{(I)}, \hat{x}^{(m+1)},\ldots,\hat{x}^{(n)}]$	
	HFM at Level II	**LFM at Level II**
Design variables	$H^{(II)}=[\hat{x}^{(m+1)},...,\hat{x}^{(l)}]$, $(l<n)$ $\hat{H}^{(I)}$ and $\hat{x}^{(l+1)},\ldots,\hat{x}^{(n)}$ are fixed	$L^{(II)}=[\hat{x}^{(m+1)},...,\hat{x}^{(n)}]$ $\hat{H}^{(I)}$ is fixed. $H^{(II)} \subseteq L^{(II)}$
Number of design variables	$l - m$	$n - m$
Optimal values	$\hat{H}^{(II)} = [\hat{x}^{(m+1)},\ldots,\hat{x}^{(l)}]$	$\hat{L}^{(II)} = [\hat{x}^{(m+1)},\ldots,\hat{x}^{(n)}]$
Corrected HFM & LFM	$X = [\hat{H}^{(I)}, \hat{H}^{(II)}, \hat{x}^{(l+1)},\ldots,\hat{x}^{(n)}]$	
	HFM at Level III	**LFM at Level III**
	Find $H^{(III)}$ and $L^{(III)}$ in a similar manner. $\hat{H}^{(I)}$ and $\hat{H}^{(II)}$ are fixed at this level.	

design variables in the first level are updated by $\hat{H}^{(I)}$, and other design variables not chosen by the HFM are also updated by $\hat{L}^{(I)}$.

The HFM chooses a different number of design variables at the next level, l, as new design variables, $H^{(II)}$, and the optimised design variables in the first level, $\hat{H}^{(I)}$, are fixed for this level. The LFM covers the whole design space except for the previous high-fidelity design spaces of $H^{(I)}$ while sharing the high-fidelity design spaces of $H^{(II)}$. The vector size of low-fidelity design variables should be $(n-m) \times 1$. The constructed multi-fidelity model finds the optimal solutions of design variables specified by $\hat{H}^{(II)}$ and $\hat{L}^{(I)}$. $\hat{H}^{(II)}$ updates the multi-fidelity model while $\hat{L}^{(I)}$ offers initial starting points for the following optimisation level. This process continues until all optimal solutions are found. It should be noted that the selection of design variables in the HFM does not follow any particular order since the LFM encompasses the whole design space regardless of choice. It was demonstrated that different selection orders do not affect the optimisation results. This modelling method can be incorporated into the probabilistic optimisation framework, where the primary challenge lies in dealing with the substantial number of high-fidelity FEM simulations.

5.2.3 *Multi-fidelity robust design optimisation method*

The developed multi-fidelity formulation is integrated with the RDO process, a type of probabilistic optimisation method. Figure 5.5 shows how the developed optimisation framework combines with the multi-fidelity method introduced in the previous section. The framework consists of the multi-fidelity modelling process and the RDO process using the concept of multi-level optimisation. The primary steps that should be highlighted in this framework are to build a set of high-fidelity design parameters at each level and construct the multi-fidelity model for the RDO process. The framework is implemented as follows.

As shown in the figure and the table, the HFM and the LFM are defined by mesh convergence studies, and then each FEM model is created using the defined mesh sizes. The number of optimisation levels, k in the figure, and the number of design variables in the HFM, m and l in the table, should be determined by engineers depending on the characteristics and size of the optimisation problem. It should be noted that the HFM has fewer design variables, $H^{(I)}$, while the LFM has all design variables in the first level, $L^{(I)}$. Once the training dataset is obtained using the two FEM models, the multi-fidelity model is constructed from the surrogate models using the

FE modelling

- Define HFM and LFM using mesh convergence studies
- Develop FEM models of both HFM and LFM

Set up sets of design parameters

- HFM having fewer design variables (*i.e.* $H^{(I)}$, $H^{(II)}$, $H^{(III)}$)
- LFM having more variables (*i.e.* $L^{(I)}$, $L^{(II)}$, $L^{(III)}$)

Select the set of $H^{(k)}$ and $L^{(k)}$

- k refers to the level of multi-level optimization

Robust design optimisation

- Monte Carlo simulations using the Sobol sampling technique
- Update HFM and LFM using optimal solutions $\hat{H}^{(k)}$ and $\hat{L}^{(k)}$

Surrogate modelling

- Build a training matrix of design points
- Obtain input-output datasets
- Create surrogate models using ANN

Multi-fidelity modelling

- Construct multi-fidelity models using the surrogate models of HFM and LFM

Complete optimisation

- Finish k-levels of optimisation
- Compare accuracy and computation cost with the conventional surrogate and multi-fidelity optimisation methods

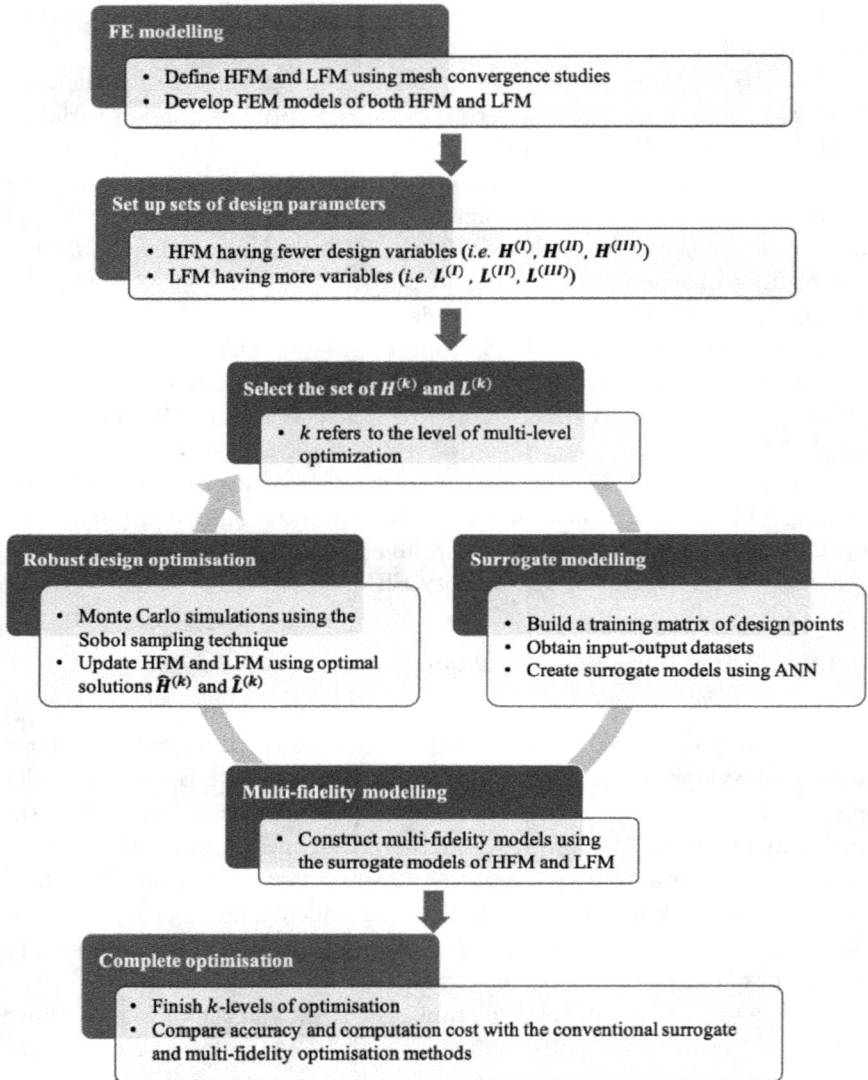

Fig. 5.5.   Multi-fidelity robust design optimisation framework.

ANN of both the HFM and the LFM based on the set of design variables, $H^{(I)}$ and $L^{(I)}$, for the first level. The multi-fidelity RDO process is carried out using Monte Carlo simulations (MCSs) employing the Sobol sampling technique and offers the optimal solutions for the first level, $\hat{H}^{(I)}$ and $\hat{L}^{(I)}$.

The optimal solutions of design variables in HFM, $\hat{H}^{(I)}$, update the HFM, while those in LFM, $\hat{L}^{(I)}$, provide starting points for the next level.

For example, there are two models in the first level of the RDO process, the HFM having only one design variable and the LFM having all design variables. Both surrogate models of the two fidelity models are constructed using the ANN. It should be noted that other design variables, apart from the selected design variable in the HFM, are fixed at the initial mean value of each design variable. At the same time, the LFM examines the design spaces of the design variables that are not selected in the HFM. Hence, both the HFM and the LFM cooperate as a multi-fidelity model during the optimisation process. The initial starting points are chosen to be evenly distributed in the design spaces through appropriate sampling techniques, such as optimal Latin hypercube sampling (OLHS) or Sobol sampling. The LFM shares the selected design variables of the HFM during the RDO process. When the first level of the RDO process is completed, the chosen design variables in the HFM are corrected by the optimal solutions of the high-fidelity design variables at this level. The optimal solutions of the low-fidelity design variables also correct the other design variables in the HFM that are not considered at the first level. These updated HFM and LFM are used in the next level of the RDO process. In this manner, the multi-fidelity optimisation process continues until the multi-fidelity model finds the optimal solutions of all high-fidelity design variables.

This new multi-level multi-fidelity modelling-based RDO framework enables the use of a considerably smaller number of HFMs compared with conventional multi-fidelity methods. These computational time savings are a result of the HFM having fewer design variables so that the number of high-fidelity FEM simulations is reduced, which is the main contribution of this multi-fidelity formulation. This method also provides the complementation of information loss using the LFM to explore the entire design space in the optimisation process, which the HFM omits depending on the selection of design variables. It should be highlighted that the proposed method takes more advantage of multi-fidelity modelling than conventional multi-fidelity methods.

5.3 Finite Element Model of Composite Structures under Nonlinear Post-Buckling

In general, stringer-stiffened composite structures are represented by a thin skin structure that should be preserved using longitudinal stringers under compressive loading conditions. The composite structures under compression are subjected to mechanical shortening. As the shortening rises along the longitudinal direction, transverse deflection appears all of a sudden at a certain shortening length or load level. This transverse deflection is called linear buckling, which was considered in Chapter 4. If a composite structure is loaded beyond the linear buckling threshold, the structure enters a post-buckling regime. Linear buckling is widely used in many industrial design problems because it conveniently provides a structural stability level for designers. However, it is acknowledged that the linear buckling load does not equate to the maximum load that the structure can withstand. Even though the applied load is several times larger than the buckling load, the failure of the structure might not occur even in the post-buckling region [97]. The post-buckling strength capacity of stringer-stiffened composite structures has been studied in the area of probabilistic design because it offers considerable benefits for reducing weight [4, 14, 16].

A nonlinear FEM model is developed in this section in order to demonstrate the multi-fidelity RDO framework using the mono-stringer stiffened composite structure under a nonlinear post-buckling regime. In the post-buckling regime, the buckled shape of the structure frequently varies as the compression load increases. Also, mode-switch or mode-jump, which represent abrupt changes in the buckling mode, is observed when the compression load increases to a specific value. These structural instabilities found in the post-buckling regime lead to significant numerical challenges, which cannot be entirely managed by the use of quasi-static FEM analysis. The nonlinear explicit dynamic analysis is a markedly superior method of analysing nonlinear post-buckling [98]. The forecast of collapse load is considerably demanding because stresses along the thickness of composites are sensitive. In general, the nonlinear post-buckling analysis of composite structures includes both geometric and material nonlinearities, which will mainly influence how the parameters of the constraints and objectives for the optimisation problem are determined. A good design approach for the mono-stringer stiffened structure aims for stiffener buckling first before the skin is in the yield region. This enables the structure to achieve buckling without damage initiation in the skin.

There are several failure measures based on the stress or strain of a structure. One well-known failure criterion, proposed by Tasi and Wu [99], can be applied to composite materials to predict the load-carrying capability of a structure. This criterion is useful to estimate the damage initiation of the structural failure process. The maximum stress failure criterion based on this theory is expressed as

$$I_{\text{Tsai–Wu}} = F_1\sigma_{11} + F_2\sigma_{22} + F_{11}\sigma_{11}^2 + F_{22}\sigma_{22}^2 + F_{66}\sigma_{12}^2 + 2F_{12}\sigma_{11}\sigma_{22} < 1.0 \tag{5.6}$$

where $I_{\text{Tsai–Wu}}$ is a failure index, F is the Tsai–Wu coefficient, σ is stress, and subscripts 1 and 2 refer to the longitudinal and transverse directions, respectively. If $I_{\text{Tsai–Wu}}$ exceeds this criterion, it is presumed that the damage begins.

Each coefficient in equation (5.6) is defined as

$$F_1 = \frac{1}{X_t} + \frac{1}{X_c}, \quad F_2 = \frac{1}{Y_t} + \frac{1}{Y_c}, \quad F_{11} = \frac{-1}{X_t X_c}, \quad F_{22} = \frac{-1}{Y_t Y_c}, \quad F_{66} = \frac{1}{S^2} \tag{5.7}$$

where X_t and X_c are the maximum tensile and compressive strength in the longitudinal direction, Y_t and Y_c are the maximum tensile and compressive strength in the transverse direction, respectively, and S is the maximum shear strength in the XY-plane.

First, the FEM model of the mono-stiffened stringer composite structure is created to examine the differences in structural behaviour depending on the use of cohesive elements. For the FEM model to account for both the geometric and material nonlinearities, the model should be created using the cohesive elements to consider progressive failures caused by interfacial debonding between the stringer and skin. Modelling these elements is a challenging task and leads to considerable computational costs due to the tiny element size in the cohesive zone. Suppose an FEM model without the cohesive elements can obtain an acceptable reaction force and failure index regarding the compressive load before global buckling is calculated using an FEM model with cohesive elements. In that case, the composite structure can be optimised geometrically in the nonlinear region. The computational cost without the cohesive elements is considerably less than that with the cohesive elements.

As shown in Figure 5.6, the same mono-stringer stiffened composite structure considered in Chapter 4 is used. This structure is clamped at both ends, but the left-hand end is free to move in the longitudinal direction (z-direction in the figure), which is the applied loading direction. A pure compression load is applied by increasing the uniform displacement at the left-hand end. The material properties are the same as in Table 4.1. The failure parameters to calculate the Tsai–Wu indexes are shown in Table 5.3.

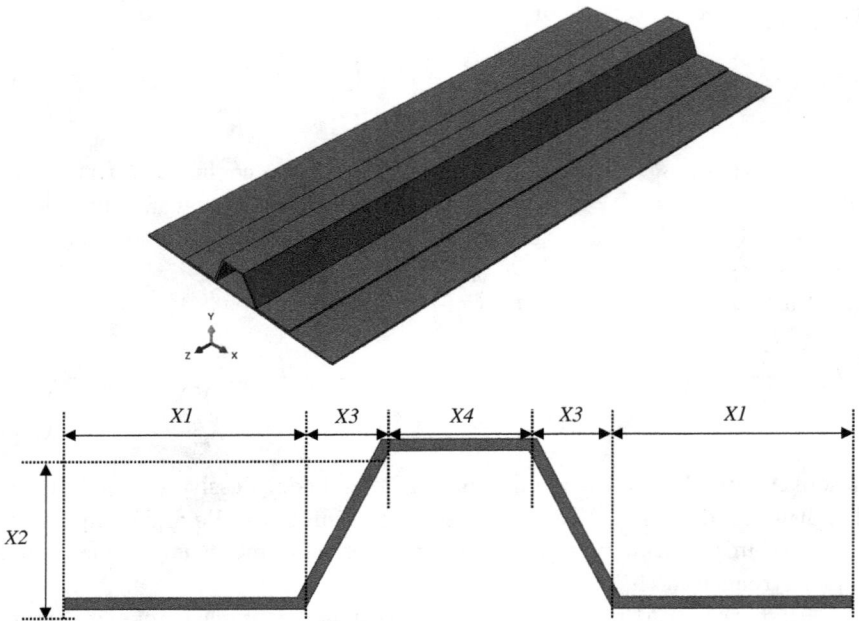

Fig. 5.6.   Mono-stringer stiffened composite panel.

Table 5.3.   Damage initiation parameters.

Parameter	Value (GPa)
Longitudinal tensile strength	2.9
Longitudinal compressive strength	1.66
Transverse tensile strength	0.058
Transverse compressive strength	0.025
In-plane shear strength	0.095

Figure 5.7 compares the reaction force between two FEM models depending on the use of the cohesive elements. The reaction force between the two different FEM models is nearly identical before global buckling occurs at a shortening length of about 3.12 mm, following linear buckling at a shortening length of about 0.5 mm. The reaction force at global buckling in the FEM model with cohesive elements was 134 kN, while the FEM model without cohesive elements resulted in 131 kN. The out-of-plane displacement of the panel at global buckling is shown in Figure 5.8. Hence, a shortening length of 3.0 mm is suitable for building up input and output

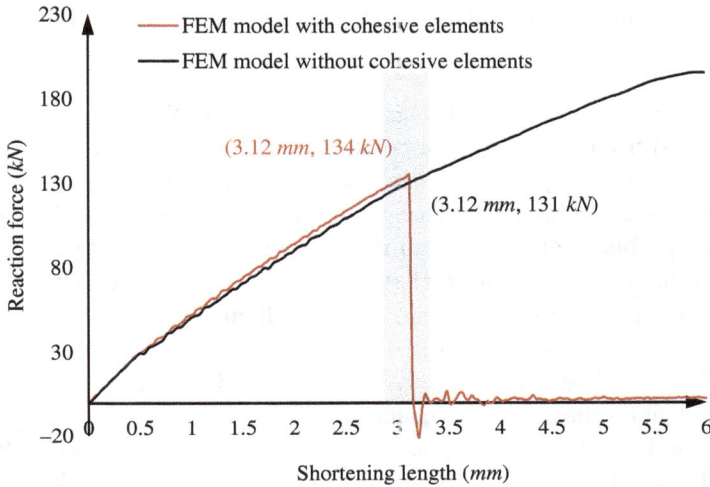

Fig. 5.7. Reaction force comparison depending on the use of the cohesive elements.

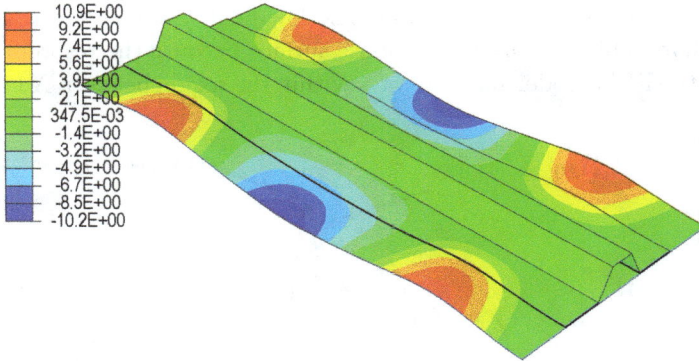

Fig. 5.8. Out-of-plane displacement at the shortening length of 3.12 mm.

datasets to construct the surrogate models. This shows that the cohesive elements need not be considered when global buckling is not a primary structural behaviour in the optimisation process.

The failure index calculated using the Tsai–Wu theory is generally considered a constraint, $I_{\text{Tsai–Wu}} < 1.0$, to track the failure process of the structure [4]. It is analysed to see whether this index can be used as a parameter that the multi-fidelity model can manage. This index is not appropriate for constructing a multi-fidelity model due to the following reasons:

- *Computation time for post-processing*: The Tsai–Wu indexes should be extracted from all integration points in all lay-ups of each element. That causes enormous post-processing time, much longer than high-fidelity FEM simulation time. For example, a single high-fidelity FEM simulation in this problem takes about 40 min; however, the computation time for post-processing takes more than 2 h and 30 min because the damage initiation point has to be found from millions of integration points in thousands of elements. This is not a reasonable computation time.
- *Local value*: The Tsai–Wu index is a local value that is considerably dependent on the mesh. It is challenging to create a surrogate model using even the HFM because it is not easy to obtain an exact reaction force at the damage initiation point where the Tsai–Wu index equals one. Even though the loading speed for the explicit dynamic analysis decreases, it is not possible to capture the acceptable reaction force using the LFM. Figure 5.9 shows the response surface of the reaction force for the LFM having one design variable ($X1$). The low-fidelity surrogate model does not thoroughly capture the response surface because the response values from the FEM model abruptly vary within the region between $X1 = 42.55$

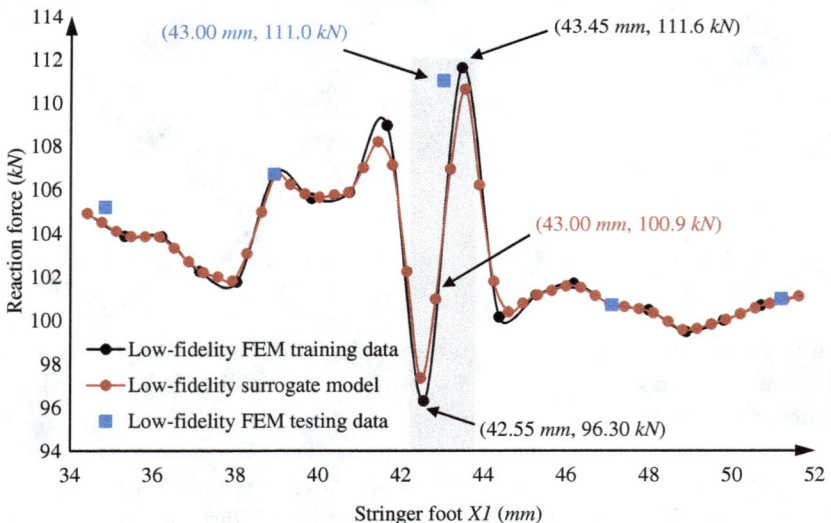

Fig. 5.9.   Response surface of the reaction force of the LFM with one design variable $X1$.

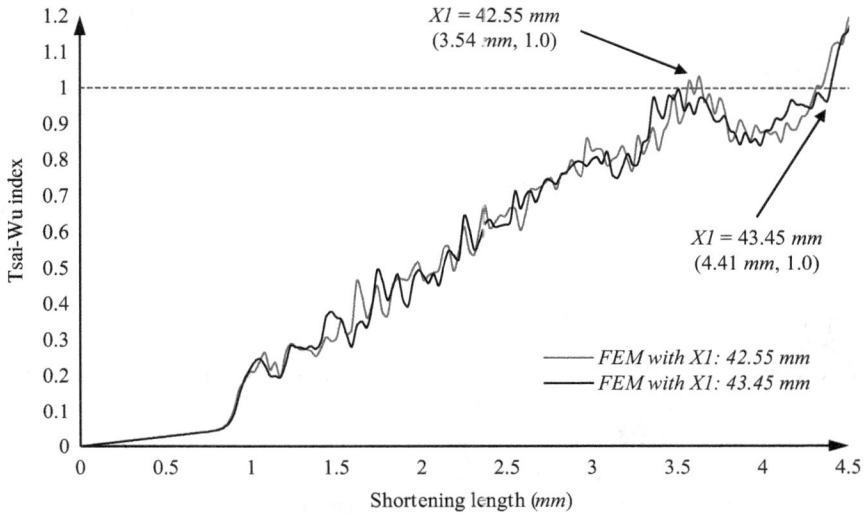

Fig. 5.10. Tsai–Wu index of low-fidelity FEM models between $X1 = 42.55$ mm and $X1 = 43.45$ mm.

mm and $X1 = 43.45$ mm. Although the surrogate model represents the responses of the FEM model except for the region, the accuracy of the surrogate model is not good enough to construct the multi-fidelity model. Figure 5.10 highlights why the Tsai–Wu index cannot be considered a parameter of the surrogate model. The figure illustrates the maximum Tsai–Wu index regarding the shortening length of two FEM models as between $X1 = 42.55$ mm and $X1 = 43.45$ mm. When the Tasi–Wu index equals one, the reaction force at the damage initiation point is obtained. However, it is challenging to collect proper reaction forces at the damage initiation point of different FEM models. In the figure, for the FEM model of 42.55 mm, the damage occurs at a shortening length of 3.54 mm. Then, the corresponding reaction force of the model is obtained. When the FEM model of 43.45 mm undergoes a similar shortening length, its reaction force is not obtained since the Tsai–Wu index is about 0.999. Then, the reaction force of the model is found at a shortening length of 4.41 mm, which leads to a sudden rise in the response surface. Hence, the reaction force at the damage initiation point is not appropriately captured, so adequate surrogate models are not obtained to construct the multi-fidelity model.

The damage initiation was not selected as a constraint based on these parameter studies; however, the reaction force was determined as an objective function. The reaction force not only represents a global index that can define the HFM and the LFM but also serve as a parameter to evaluate global buckling. Hence, material properties were assumed in the linear elastic region, while geometric nonlinearity was considered. Delamination, which is caused by debonding between the stringer and the skin, was not considered in this section.

5.4 Numerical Examples

Two engineering approaches, DO and RDO, were carried out to design a mono-stringer stiffened composite structure using the developed multi-level multi-fidelity probabilistic optimisation method. First, DO, which is more straightforward than probabilistic optimisation methods, was conducted to evaluate the feasibility of the developed method. Then, RDO was carried out to see whether this new method could be systematically incorporated into the probabilistic optimisation of composite structures considering design uncertainties. In these two engineering examples, the efficiency of the multi-fidelity model covering different design spaces between the HFM and the LFM was evaluated using the conventional optimisation method based on high-fidelity modelling-based surrogate models.

5.4.1 *Composite structures under nonlinear post-buckling*

As shown in Figure 5.6, we use the same mono-stringer stiffened composite structure considered in the example in Chapter 4. It should be noted that only the geometry of the stringer part is to be optimised, and the dimensions of the skin are fixed during the optimisation process. Four stringer geometry parameters ($X1$, $X2$, $X3$, and $X4$) are considered to validate the proposed method via a simple optimisation problem. The type of fidelity in these two examples was chosen by the level of discretisation of the FEM model. Figure 5.11 shows that size of the mesh grid of HFM and LFM was 7.0 and 20.0 mm, respectively. The HFM and LFM show about a 10% difference in accuracy, while the computational cost of the HFM is 10 times higher than that of the LFM. These models were modelled using nonlinear explicit dynamic finite element analysis using Abaqus/CAE [66]. The FEM models consist of four-node shell elements (S4R).

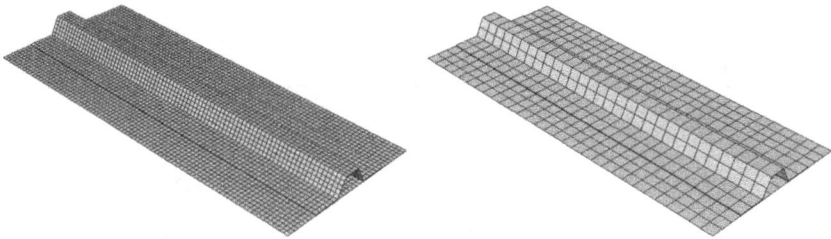

Fig. 5.11. The HFM with 7 mm mesh size (left) and the LFM with 20 mm mesh size (right).

Two types of examples, i.e., DO and RDO, are presented to demonstrate the developed multi-fidelity modelling-based probabilistic optimisation method embracing different design spaces between the HFM and the LFM. Each level has the same objective functions and constraints, whereas other design variables are considered at each level. The boundary conditions and loading conditions were not changed at every optimisation level. The optimisation aims to find a composite panel that shows minimum mass and maximum reaction force under the nonlinear post-buckling regime. Through these two engineering examples, the potential application area of the developed optimisation method can be broadened to large-scale problems having substantially larger design spaces.

5.4.2 *Multi-fidelity deterministic optimisation*

DO was conducted to see if this method provides acceptable solution accuracy and computational time savings. In order to find the best solution, the Pareto front is found first, and then the optimum solution is selected by a decision-maker.

5.4.2.1 *Problem definition*

As can be seen in Table 5.4, the mesh size of the HFM and LFM for finite element analysis was defined to be 7.0 and 20.0 mm, respectively. The design spaces of cross-section geometric parameters of the stringer are also denoted. The optimisation was carried out using the NSGA-II algorithm, a multi-objective method suited well for highly nonlinear design spaces. The Pareto front is constructed by choosing the feasible non-dominated designs, where each design point has the best combination of objective function

Table 5.4.   DO: problem definition.

Description		Value
Multi-fidelity model	HFM	*mesh*: 7 mm
	LFM	*mesh*: 20 mm
Design variables	Stringer foot	$34.4 \leq X1 \leq 51.6$
	Stringer height	$24.0 \leq X2 \leq 36.0$
	Distance between top and foot	$12.0 \leq X3 \leq 18.0$
	Stringer top	$25.0 \leq X4 \leq 30.0$
Optimisation method	NSGA-II	*generation*: 20
		population: 12
Constraints	HFM Reaction Force (RF)	$RF \geq 124.0$ kN
	LFM RF	$RF \geq 130.0$ kN
Objectives	HFM Mass	*minimise*
	LFM Mass	*minimise*
	HFM RF	*maximise*
	LFM RF	*maximise*

values. The improvement of the objective is only possible by sacrificing the other objectives. Generation and population numbers for the optimisation process were chosen as 20 and 12, respectively. These values were obtained from the convergence study of the objective functions. Hence, the total number of FEM simulations at each level was 240. Different constraints for the HFM and the LFM were defined by the minimum reaction force that originates from the initial model. Both the HFM and the LFM have their objectives in the multi-fidelity modelling-based DO process.

Table 5.5 shows the selected design variables in the HFM at each level and the number of FEM simulations to build the multi-fidelity model. It is not surprising that the number of high-fidelity FEM simulations is dramatically reduced because the LFM examines the whole design space. This should compensate for the disparity in the solution space caused by a lack of information resulting from the fact that the HFM does not cover the entire design space.

5.4.2.2 *Results*

There are four objectives in this optimisation problem, and this is not easy to present using a 2D or 3D graph. However, a tabular form is an option

Table 5.5. Details of the multi-fidelity model.

| | HFM | | LFM | |
| | Design variable | Number of FEM simulations | Design variable | Number of FEM simulations |
Level				
I	$X1$	10	$X1, X2, X3, X4$	100
II	$X2$	10	$X2, X3, X4$	70
III	$X3$	10	$X3, X4$	40
IV	$X4$	10		—

Table 5.6. DO results: multi-fidelity model.

| | Design variable | | Result | | | | Optimal value (mm) for multi-fidelity model correction | | HFM updated by multi-fidelity model |
| | | | RF (kN) | | Mass (g) | | Value from HFM | Value from LFM | |
Level	HFM	LFM	HFM	LFM	HFM	LFM			
I	$X1$	$X1$	126	138	617	635	$X1$: 47.225	$X2$: 34.470	$X1$: 47.225 (Fixed)
		$X2$						$X3$: 17.450	$X3$: 17.450
		$X3$						$X4$: 27.515	$X4$: 27.515
		$X4$							
II	$X2$	$X2$	132	141	639	641	$X2$: 35.924	$X3$: 17.241	$X2$: 35.924 (Fixed)
		$X3$						$X4$: 29.728	$X4$: 29.728
		$X4$							
III	$X3$	$X3$	133	140	642	641	$X3$: 17.493	$X4$: 29.282	$X3$: 17.493 (Fixed)
		$X4$							
IV	$X4$	—	133	—	642	—	$X4$: 30.000	—	$X4$: 30.000 (Fixed)

to represent results concerning different levels and objectives. Table 5.6 shows the results of DO using the multi-fidelity model. This model has four geometric parameters to be optimised; hence, the HFM has only one design variable at each optimisation level in this example. Table 5.6 highlights how both the HFM and the LFM embrace the whole design space as a multi-fidelity model. As mentioned before, two fidelity models share the design variable of $X1$ during the optimisation process at Level I. The results of Level I show optimal values of both the HFM and the LFM that provide the maximum reaction force at the given shortening length. These optimal values correct both the HFM and the LFM; in particular, the optimal value of $X1$ is fixed because this value is the best solution obtained from the high-fidelity design spaces. At Level II, the HFM takes the new design variable $X2$. At the same time, $X3$ and $X4$ are also updated by the optimal values of the LFM at Level II. The LFM has three design variables ($X2$, $X3$, and $X4$), with $X1$ having a fixed value from Level I. Similarly, Level II provides the optimal values, namely $X2$ of the HFM and $X3$ and $X4$ of the LFM. They are also used to correct both the HFM and the LFM in Level II, and these updated models are utilised at Level III. In this manner, the final optimal solution is found when Level IV is completed.

Figure 5.12 shows the Pareto front of the optimisation using the multi-fidelity model. As shown in the figure, the Pareto front improves gradually in the direction in which the reaction force increases as the level goes up. The Pareto front of Level I is lower than that of Level II. The Pareto front of Level IV is higher than that of the other levels. Table 5.7 shows that the maximum reaction force of each level increased from Level I (126.0 kN) to Level IV (133.1 kN). The conventional DO method using high-fidelity

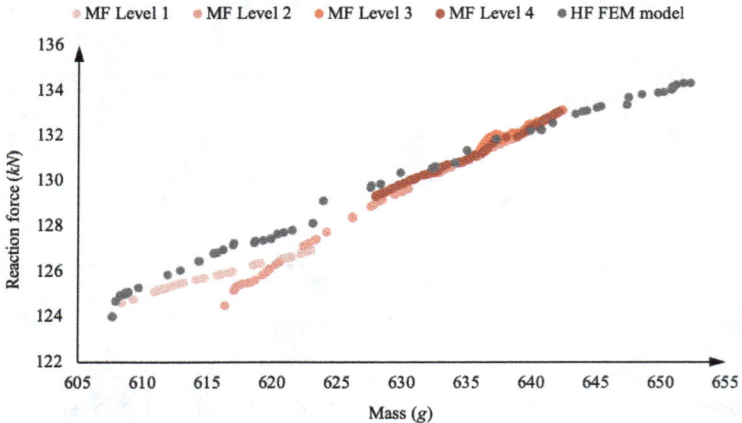

Fig. 5.12. DO results comparison by the Pareto front: multi-fidelity vs. high-fidelity FEM.

Table 5.7. DO results accuracy by optimal solution: multi-fidelity vs. high-fidelity.

	Multi-fidelity model	High-fidelity FEM model
Maximum RF (kN)	133.1	132.8
Minimum mass (g)	642	640

FEM models has the same objectives and constraints, and the Pareto front of this traditional method is shown in Figure 5.12. Table 5.7 presents a comparison of the final chosen solution between the multi-fidelity model and the high-fidelity FEM model to show how accurate the solution of the multi-fidelity model is. The reaction force of the multi-fidelity model indicates good agreement with that of the high-fidelity FEM model, corresponding to a mass of 640 g.

The proposed multi-fidelity modelling method aims to provide the accuracy of solutions and computation time savings. It is significant to show how much computational time savings are obtained using the multi-fidelity model. Unfortunately, not many research works provide a standard guideline for the computational cost of multi-fidelity models [29]. In general, the computational cost is defined by the number of high-fidelity FEM simulations because the primary goal of the multi-fidelity modelling method is to reduce the computational cost of the HFM using the LFM [57]. In order to evaluate the computational cost of each model and diminish computational noise, the total number of FEM simulations to create the multi-fidelity model is calculated, and then it is normalised by the number of FEM simulations used for the conventional method [32]. Figure 5.13 displays the computational time taken by the proposed method and the

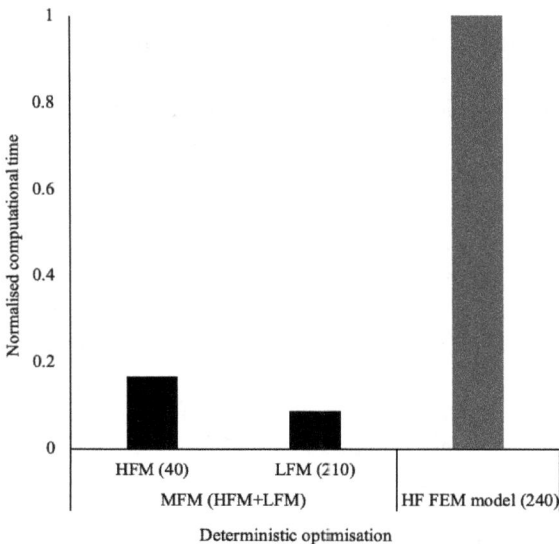

Fig. 5.13.   DO computation time comparison: multi-fidelity vs. high-fidelity.

traditional method. In the previous section, 40 design points of the HFM and 210 design points of the LFM were used to construct the multi-fidelity model. The conventional method was performed using 240 high-fidelity FEM models. The developed multi-fidelity model offers about 80% of computation time savings compared to the traditional method using the high-fidelity FEM models.

5.4.3 *Multi-fidelity robust design optimisation*

Through the benchmark study of DO in the previous section, the feasibility of this developed multi-fidelity modelling-based optimisation method was proven. This enables us to broaden to the area of RDO using the developed method. This example aims to demonstrate how computationally economical and accurate this proposed optimisation method is to find a robust design solution.

5.4.3.1 *Problem definition*

Table 5.8 shows the problem definition of this example. The details of the stiffened composite panel, the mesh size, and the design space are the same as the previous example. The reason for the same design space with DO is that a robust solution might be far from the deterministic solution. Each design variable has its own design uncertainty of 0.1% as a manufacturing tolerance [16]. These uncertainties are described by the statistical characteristics, which are in the form of a normal distribution having a mean and standard deviation. The extra constraint was added by a maximum mass of 630 g. In particular, the objectives in Table 5.8 present the features of RDO. The eight objectives for the optimisation process using the multi-fidelity model are to maximise reaction force while minimising mass and the standard deviations of both reaction force and mass. Table 5.5 shows the selection order of the design variable in the HFM at each level and the number of FEM simulations to create the multi-fidelity model. Figure 5.14 illustrates the multi-fidelity RDO framework using NSGA-II. In particular, robustness should be checked at each member of the population of each generation. MCS was utilised to specify the statistical moments of the objectives (mass and reaction force), which are caused by the uncertainties of random design variables. The Sobol sampling technique was considered, which provides more uniformly distributed design points and more robust statistical predictions than the descriptive sampling

Table 5.8. RDO: problem definition.

Description		Value
Multi-fidelity model	HFM	*mesh*: 7.0 mm
	LFM	*mesh*: 20.0 mm
Design variables	Stringer foot	$34.4 \leq X1 \leq 51.6$
	Stringer height	$24.0 \leq X2 \leq 36.0$
	Distance between top and foot	$12.0 \leq X3 \leq 18.0$
	Stringer top	$25.0 \leq X4 \leq 30.0$
Optimisation method	NSGA-II	*generation*: 10
		population: 12
Analysis type	Monte Carlo simulation	*Sobol sampling*: 1000
Design uncertainty	Mean	Standard deviation (Std. dev.)
	$X1$	$0.001 \times X1$
	$X2$	$0.001 \times X2$
	$X3$	$0.001 \times X3$
	$X4$	$0.001 \times X4$
Constraints	HFM Reaction Force (RF)	$RF \geq 124.0$ kN
	LFM RF	$RF \geq 130.0$ kN
	Mass	$mass \leq 124.0$ kN
Objectives	HFM Mass	*minimise*
	LFM Mass	*minimise*
	Std. dev. HFM Mass	*minimise*
	Std. dev. LFM Mass	*minimise*
	HFM RF	*maximise*
	LFM RF	*maximise*
	Std. dev. HFM RF	*minimise*
	Std. dev. LFM RF	*minimise*

technique [58]. The maximum number of sampling points to check the robustness was 1,000. The convergence tolerance was 0.1% of both mean and standard deviation compared with those associated values calculated every 25 sampled points. Hence, the maximum number of the multi-fidelity model simulations was 120,000 during the robust optimisation process.

5.4.3.2 *Results*

Table 5.9 describes the optimisation results at each level and how the HFM is corrected by the optimal values from using the multi-fidelity model. In the

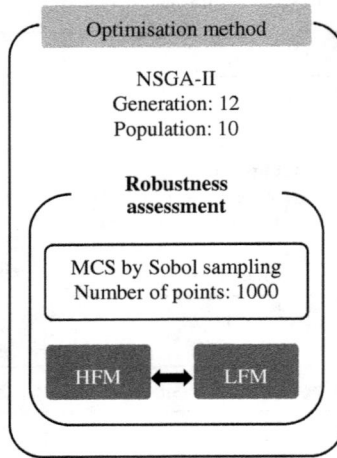

Fig. 5.14.   RDO structure.

Table 5.9.   RDO results: multi-fidelity model.

Level	Design variable		RF (kN)		Mass (g)		Optimal value (mm) for multi-fidelity model correction		HFM updated by multi-fidelity model
	HFM	LFM	HFM	LFM	HFM	LFM	Value from HFM	Value from LFM	
I	X1	X1	127	133	623	627	X1: 50.129	X2: 31.589	X1: 50.129 (Fixed)
		X2						X3: 17.684	X3: 17.684
		X3						X4: 20.656	X4: 20.656
		X4							
II	X2	X2	129	131	631	617	X2: 33.312	X3: 15.082	X2: 33.312 (Fixed)
		X3						X4: 20.097	X4: 20.097
		X4							
III	X3	X3	129	135	630	625	X3: 18.000	X4: 23.640	X3: 18.000 (Fixed)
		X4							
IV	X4	—	129	—	631	—	X4: 20.504	—	X4: 20.504 (Fixed)

first level, both the HFM and the LFM share the design variable of X1 while exploring the entire solution spaces of all design variables using the LFM. Figure 5.15 depicts how this multi-fidelity modelling-based method probes the solution spaces when the HFM and the LFM have a different number

HFM
(*X1*)

LFM
(*X1, X2, X3, X4*)

3-1

(b)

Level
I

(a)

(c)

(d)

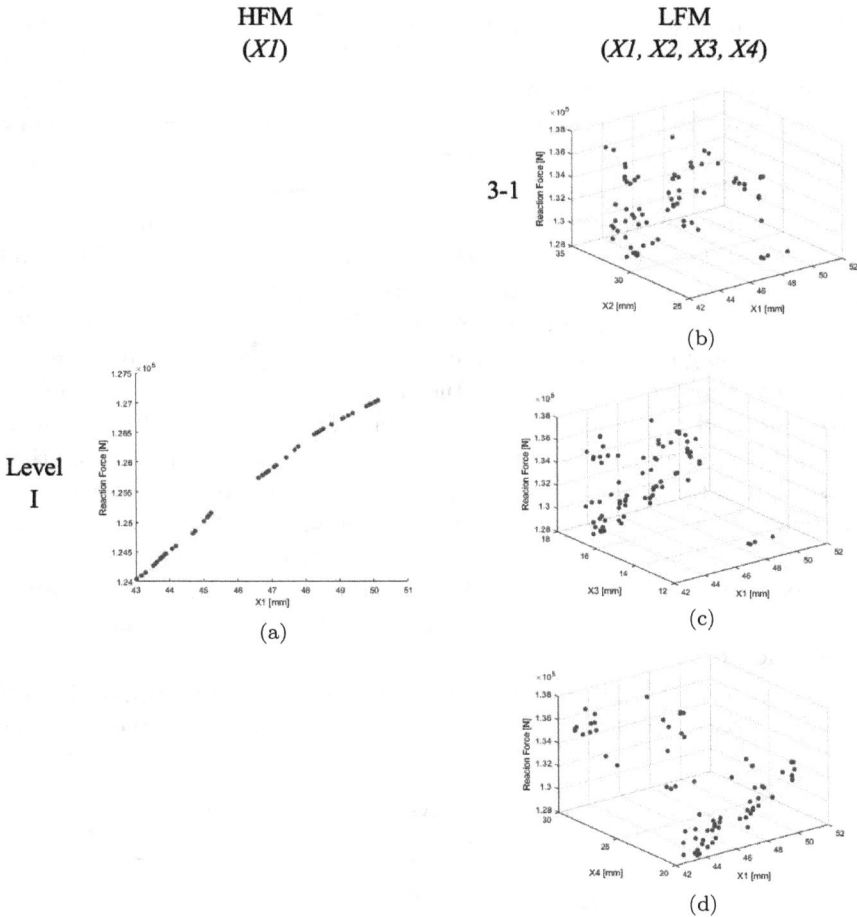

Fig. 5.15. Multi-fidelity modelling-based optimisation at Level I.

of design variables in the optimisation process. Figure 5.15(a) shows the optimal solution of X1 using the HFM, which is reliable and accurate. At the same time, Figures 5.15(b)–5.15(d) represent how the proposed multi-fidelity modelling method is implemented to embrace different solution spaces which are not included in the HFM during the optimisation process. It should be highlighted that these subfigures offer what the different solution spaces are expected by the use of the LFM sharing the same X1 with the HFM at the first level. As can be seen in the figures, the LFM helps the decision-maker explore the different solution spaces of $X2$, $X3$, and $X4$, corresponding to the variation of $X1$ of the HFM. This information trade within the multi-fidelity model enables $X2$, $X3$, and $X4$ of the HFM to be updated by optimal values of the LFM having all design variables. Table 5.9 highlights that the reaction force of the HFM at the given shortening length at each level rises gradually as the optimisation level is escalated.

Figure 5.16 illustrates the Pareto front at each level of the RDO process using the multi-fidelity model. As seen in the Pareto front in the figure, the reaction force in the post-buckling regime was improved as the level progressed, while the constraint of mass was not violated. Finally, the Pareto front of the fourth level presents the maximum reaction force with a mass of 630 g to evaluate the efficiency of the proposed method using an equivalent approach. In the figure, the optimal solutions from the use of the surrogate model based on 200 design points of the HFM are also presented to verify the accuracy of the multi-fidelity model. It should be noted that these Pareto fronts from both the proposed method and the conventional method show an acceptable range of difference among the optimal solutions.

Table 5.10 describes the robust results which are chosen by the optimal solutions. This RDO aims to maximise the mean values of mass and reaction force and to minimise the standard deviation values of mass and reaction force. Hence, two optimal robust designs were selected that show the maximum mean of reaction force and the minimum standard deviation of reaction force, respectively. These results were compared with the results of

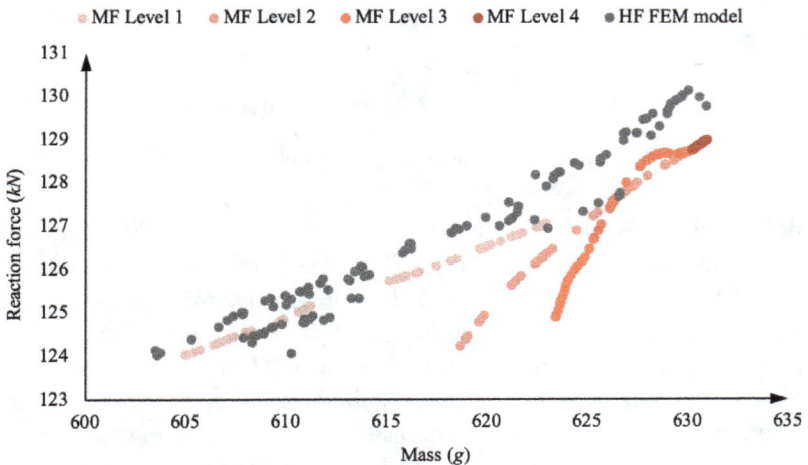

Fig. 5.16.   RDO result comparison by the Pareto front: multi-fidelity vs. high-fidelity.

Table 5.10.   RDO results' accuracy by optimal solution: multi-fidelity vs. high-fidelity.

Model	X1 (mm)	X2 (mm)	X3 (mm)	X4 (mm)	RF (kN) Mean	Std. dev.	Mass (g) Mean	Std. dev.
Objective: maximise reaction force								
Multi-fidelity model	50.13	33.3	18	20.5	129	0.3	631	0.12
HF surrogate model	43.47	35.5	17.3	30	130	0.48	631	0.16
Objective: minimise standard deviation of reaction force								
Multi-fidelity model	43.01	32.7	15.4	22.3	125	0.18	608	0.15
HF surrogate model	40.87	35.4	14	24.4	125	0.24	611	0.15

the conventional high-fidelity surrogate model that have the same mass as those designs from the proposed method. The robust solution of the multi-fidelity model is close enough to that of the high-fidelity surrogate model. Both solutions for different objectives have an acceptable agreement with the solutions of the high-fidelity surrogate model. It should be highlighted that the optimal solutions of the multi-fidelity model are more robust, in which the standard deviations of each reaction force are 0.30 and 0.18, respectively, compared to those of the high-fidelity surrogate model. This means that the proposed optimisation method using a much smaller number of FEM simulations discovers a more robust design. This is also found for the standard deviation of mass.

In general, surrogate models are used to conduct RDO to overcome high computational costs, and these models are a kind of black-box form between design variables and responses. Hence, the final chosen solution that is produced by these surrogate models should be validated using proper calculation, such as an FEM solver or an experiment. Table 5.11 shows the solution accuracy of the surrogate model constructed using the ANN. According to the table, the reaction forces of two surrogate models at the optimal design point are nearly identical to those calculated using the FEM solver. It should be remarked that the accuracy of the multi-fidelity model is better than that of the HFM.

Figure 5.17 shows the statistical characteristic of the reaction force depending on the type of optimisation. The optimal design, which is obtained by DO, represents a more significant mean value of reaction

Table 5.11.   The accuracy of the surrogate model: ANN vs. FEM solver.

Model	$X1$ (mm)	$X2$ (mm)	$X3$ (mm)	$X4$ (mm)	RF (kN) (ANN)	RF (kN) (FEM)	Error (%)
Multi-fidelity model	50.1	33.3	18.0	20.5	129.0	129.0	0%
HF surrogate model	43.5	35.5	17.3	30.0	130.1	128.2	1.50%

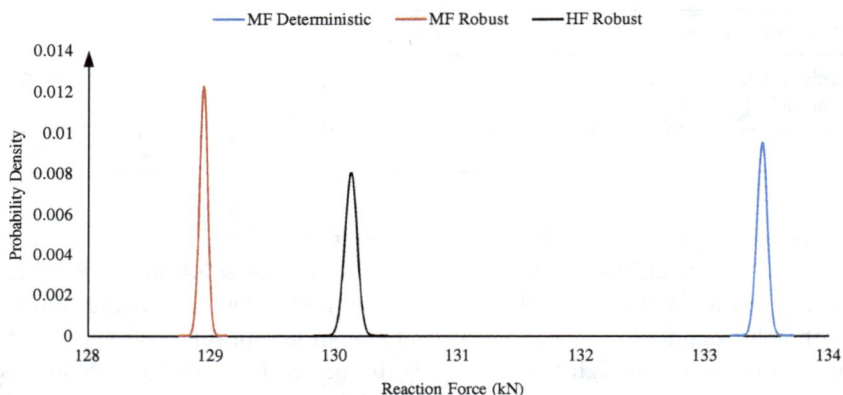

Fig. 5.17.   Statistical comparison between DO and RDO: maximum reaction force.

force than the robust design. In contrast, the deterministic design is more sensitive, with a larger standard deviation. It is worth noting that the variation of the robust design using the multi-fidelity model is smaller than that of the conventional method. Figure 5.18 illustrates the robust designs of the multi-fidelity model and the high-fidelity surrogate model when the objective function minimises the standard deviation of the reaction force under the post-buckling regime. Both robust designs in the figure have a nearly identical mean value of reaction force, while the dispersion of reaction force obtained by the multi-fidelity model is smaller than that of the conventional method. This is also clearly observed in Table 5.10. It should be noted that the accuracy of this multi-fidelity model is good enough to be applied to RDO considering the design uncertainty of random design variables.

The computational time savings are presented in Figure 5.19. The computational cost to create the multi-fidelity model was normalised by the total computation time of the high-fidelity surrogate model. The developed

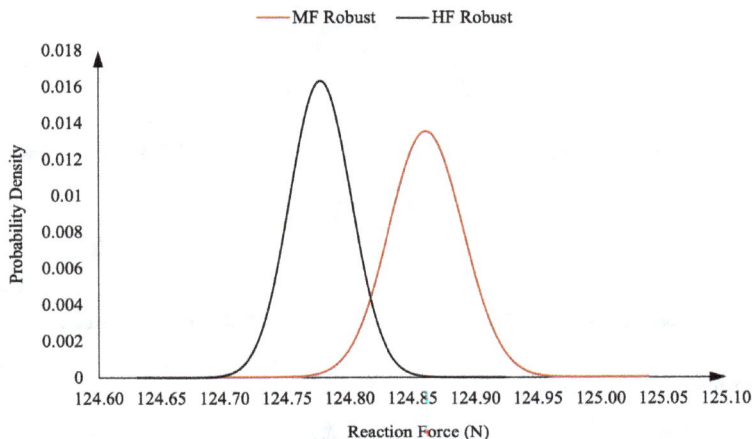

Fig. 5.18. Statistical comparison between multi-fidelity and high-fidelity surrogate model: minimum standard deviation of reaction force.

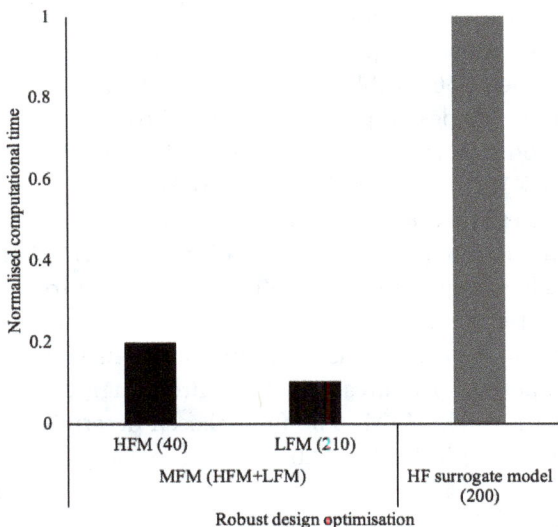

Fig. 5.19. RDO computation time comparison: multi-fidelity vs. high-fidelity.

multi-fidelity model incorporated 40 design points of the HFM and 210 design points of the LFM, while the high-fidelity surrogate model used 200 design points of the HFM. The computation time savings through the use of the multi-fidelity model was about 70% compared to the computation

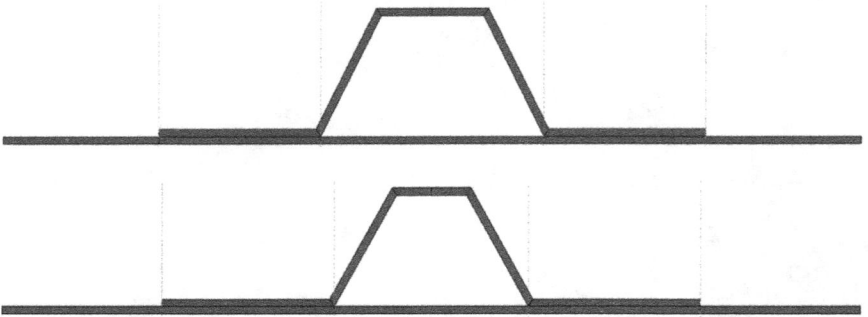

Fig. 5.20. Chosen stringer geometry from DO (top) and RDO (bottom): maximum reaction force.

time of the conventional high-fidelity surrogate method. It is critical to show how much the proposed multi-fidelity method is efficient compared to the different multi-fidelity methods covering the same number of design variables between the HFM and the LFM. These conventional multi-fidelity methods require the same number of FEM simulations between the HFM and the LFM to construct proper response correction surfaces that are the main component [56]. Table 5.5 shows that the LFM having all design variables requires 100 design points of the LFM to construct the surrogate model with good quality. Hence, the HFM having all design variables should also need 100 design points of the HFM to create proper response correction surfaces. Figure 5.19 shows that the multi-fidelity model using the proposed method requires 40 design points of the HFM and 210 points of the LFM. The computational time savings of the proposed method is about 50% greater compared to the conventional multi-fidelity methods. These savings could be a dramatic improvement in conducting RDO of large-scale composite structures, which require a high computational cost to analyse even a single HFM. The final chosen deterministic and robust designs are illustrated in Figure 5.20.

5.5 Summary

In this chapter, the multi-level multi-fidelity modelling-based optimisation framework has been presented. It was demonstrated by the RDO of a mono-stringer stiffened composite panel under the nonlinear post-buckling regime. It should be noted that the HFM has fewer design variables, whereas the LFM has more design variables in this multi-fidelity modelling method. This is one of the main contributions of this work compared to conventional multi-fidelity modelling methods that require the same number of design variables between the HFM and the LFM to create the correction response function. During this optimisation process, the HFM with fewer design variables provides an accurate solution to correct the multi-fidelity model. At the same time, the LFM with more design variables explores the solution space of all design variables while sharing the design variables with the HFM. In particular, the constructed multi-fidelity model using an ANN is incorporated with the multi-level optimisation approach to deal with large-scale composite problems with many design spaces. The results of two engineering examples were evaluated in terms of solution accuracy and computational time savings. The optimal solutions of DO and RDO using the multi-fidelity model were nearly identical to those using the conventional method. It should be highlighted that the standard deviation of the optimal solution using the multi-fidelity model was more robust than that of the optimal solution using the conventional high-fidelity surrogate model. Computational efficiency was highlighted by the use of normalisation between the number of FEM simulations to create the multi-fidelity model and the number of FEM simulations to generate conventional high-fidelity surrogate models, as well as a comparison with a traditional multi-fidelity model with the same number of design variables for the HFM and the LFM. The multi-fidelity model was constructed using both 40 design points of the HFM and 210 design points of the LFM. In contrast, the conventional method for DO and RDO used 240 and 200 design points of the HFM, respectively. At least 50% of computational time savings were obtained using this new multi-level multi-fidelity modelling method. Through these demonstrations, the multi-fidelity modelling-based optimisation framework was proven as a new optimisation method that the HFM and the LFM have different design spaces during the optimisation process. The developed multi-fidelity modelling-based probabilistic optimisation framework shows excellent potential for large-scale composite problems considering design uncertainties.

Chapter 6

Multi-Fidelity Probabilistic Optimisation Using Sparse High-Fidelity Information

Thermo-mechanical loading may lead to a premature structural collapse in composite materials used in different engineering structures. By comprehensive awareness of the thermo-mechanical loading behind the structural stability, the stability of a designed composite system can be more precisely secured. As industrial demands for energy efficiency and low carbon emissions are rising, structures using these composite materials are being applied to a broad range of engineering fields due to their fundamental advantages of high strength and light weight. For example, composite structures are used in aircraft structural design because high-speed and lightweight aircraft requirements are significantly growing. Lightweight structures are generally vulnerable to buckling under extreme environmental conditions. Hence, it is essential that buckling is taken into account at the design stage of composite structures. In particular, thermo-mechanical buckling should not be ignored since thermal loading can cause considerable damage followed by mechanical loading for a high-speed aircraft. Many design optimisation approaches for composite structures have been developed to consider both thermal and mechanical loading. However, the limitation of these current approaches is that they are only based on deterministic optimisation (DO), which could result in conservative composite designs because they do not consider the design uncertainties associated with the product's lifecycle.

Commonly, these uncertainties are considered by the use of safety factors that could result in a waste of material resources caused by increased weight. An ideal optimisation approach in considering the design uncertainties is probabilistic design optimisation, including reliability-based design optimisation (RBDO) and robust design optimisation (RDO). These optimisation approaches allow structural optimisation to deliver low carbon emissions and energy efficiency since they yield reliable and robust design solutions for the entire lifecycle of structures. Specifically, RBDO aims to minimise the probability of failure so that the final design satisfies reliability requirements based on the probabilistic characteristics of the design variables. This optimisation approach focuses on the safety of structures when they are exposed to catastrophic and extreme circumstances. So far, the vast majority of research works have considered the thermal buckling behaviours of composite and metallic structures caused by thermal expansions without considering mechanical loading [100–104]. Only a few research works have studied the thermo-mechanical buckling behaviours of composite structures. They have been demonstrated by structural analysis and deterministic design problems that do not consider the design uncertainties [18, 20]. Although probabilistic design optimisation has been studied for designing different composite structures [4, 15, 16, 24, 32, 51], there has not been any research work on the probabilistic design of composite structures under thermo-mechanical loading. It is not surprising that the computational cost of these design problems is higher than that of single-loading cases.

In the design area of composite structures, RBDO considers different design uncertainties depending on the objectives of optimisation problems. It highlights how different reliable designs rely on the consideration of design uncertainties, in contrast to deterministic designs. These design uncertainties commonly involve mechanical properties and geometric parameters of structures associated with the design and manufacturing process [15, 16, 76, 105–107]. As discussed in previous chapters, the probabilistic design optimisation process encounters significant computational challenges due to statistical calculations regarding design uncertainties. In particular, RBDO includes reliability assessments to obtain a prescribed reliability level, which requires millions of computationally expensive simulations such as finite element method (FEM) simulations. An attempt to reduce the high computational cost resulting from considering design uncertainties involves the use of surrogate models [4, 10, 57]. These models have been widely used in various probabilistic optimisation (PO) processes for composite structures while providing a certain level of computational gains. However, the computational cost is still the main obstacle to carrying out the RBDO of complex and large-scale composite structures. It is not surprising that even a single FEM simulation in these problems is too computationally expensive to create surrogate models.

A technique that addresses this prohibitive computational challenge involves the use of a multi-fidelity modelling approach that offers substantial computational time savings compared with the traditional single-fidelity surrogate modelling approach. This multi-fidelity modelling approach allows hundreds of computationally economical low-fidelity model (LFM) data points to boost the accuracy of tens of high-fidelity model (HFM) data points, accurate but computationally demanding. In particular, the multi-fidelity modelling approach enables the probabilistic optimisation to account for the design uncertainties that conventional surrogate modelling approaches are still struggling with [13, 29, 57]. As introduced in Chapters 4 and 5, traditional multi-fidelity approaches using artificial neural networks (ANNs) or the response surface method (RSM) require that the HFM and LFM carry the same number of training points to build correction response surfaces [32, 56]. This may lead to additional high-fidelity FEM simulations that designers would rather avoid. In contrast, some multi-fidelity approaches allow the HFM and the LFM to embrace a different number of training points since they don't build correction response surfaces. They are mainly derived using Gaussian processes (GPs) [63], and these approaches are combined with the linear autoregressive information fusion scheme [72, 73]. One notable example is multi-fidelity optimisation using co-kriging, which was developed to consider the linear correlation between different fidelity levels using GPs [72]. The authors demonstrated the multi-fidelity method using composite design problems in which the fidelity is defined by the linearity level of the numerical solver. Another notable example is a nonlinear information fusion algorithm presented to capture nonlinear correlations during the multi-fidelity modelling process [108, 109]. The algorithm was based on GP regression and the nonlinear autoregressive scheme to consider the complex nonlinear correlation between different fidelity models. The algorithm offered a much higher accuracy when the two fidelity models have nonlinear correlations across the whole design space. This can contribute to dealing with a challenging gap in the response surface caused by different FEM mesh sizes between the HFM and the LFM. This gap is not captured even if hundreds of LFM data points are used when traditional multi-fidelity approaches carry out structural optimisation problems. Due to their many advantages, GPs considering linear correlation have found extensive use with different optimisation problems, with notable examples being Refs. [72, 84]. However, their application to structural optimisation problems considering the nonlinear information fusion algorithm has not received attention.

The multi-fidelity modelling approaches for probabilistic optimisation developed so far require that the HFM and LFM explore the same dimension of design spaces. This causes enormous computational cost when there are many design variables in large-scale composite design problems. As presented in Chapter 5, the multi-level multi-fidelity modelling method allows two different fidelity models to embrace different design spaces during the optimisation process. This method is not suitable for probabilistic design problems that show a nonlinear correlation between the HFM and the LFM. The method does not use GPs considering the correlation between the two models but uses an ANN for the analysis of input–output values to construct the multi-fidelity model. This drawback could be resolved by the use of both GPs and the nonlinear information fusion algorithm. The improved multi-fidelity method could enable the LFM to represent the response surfaces using the HFM covering part of the entire design space. This research work aims to develop, for the first time, a method for the multi-fidelity probabilistic optimisation of composite structures under thermo-mechanical loading using GPs. Prior to the work presented in this monograph, no work has been carried out by the research community on the topic of the multi-fidelity probabilistic optimisation of composite structures under thermo-mechanical loading.

In summary, the work presented in this chapter aims to develop a novel multi-fidelity methodology for:

- Formulating a multi-fidelity method. The multi-fidelity formulation allows the HFM and LFM to have a different number of design variables during the optimisation process. Specifically, the HFM supervises only a small part of the entire design space, while the LFM encompasses the whole design space. The primary contribution of this formulation allows the high-fidelity training dataset to be collected by different sampling levels, such as dense and sparse. This offers a very efficient means of constructing a multi-fidelity model.
- Combining the nonlinear information fusion algorithm with the multi-fidelity modelling formulation. Previous works on this topic have exclusively involved the use of high-fidelity information covering the same dimension of design space as low-fidelity information. One drawback of this concept is that it causes extra computational cost. The proposed multi-fidelity probabilistic framework avoids this problem by supervising

part of the whole design space using densely distributed high-fidelity information while covering the entire design space using sparsely distributed information. Then, the nonlinear information fusion algorithm using GPs calculates the correlation between the HFM and the LFM and incorporates the multi-fidelity RBDO framework to construct a more accurate multi-fidelity model.

- Demonstrating a novel multi-fidelity probabilistic optimisation method. As part of this approach, a new multi-fidelity methodology is developed. This methodology is demonstrated by the RBDO of composite structures under thermo-mechanical loading for the first time. The accuracy and computational efficiency are evaluated through this numerical example when the composite structure is subjected to mechanical loading and thermal loading.

This chapter begins by reviewing the fundamental theory of thermo-mechanical buckling and the nonlinear information fusion algorithm. Then, the developed multi-fidelity method is introduced, including the multi-fidelity formulation and RBDO framework. A numerical example of a composite structure under thermo-mechanical loading is presented towards the end of this chapter. The work presented in this chapter is based on the authors' work [110].

6.1 Multi-Fidelity Probabilistic Optimisation

A novel multi-fidelity modelling method, which blends an HFM and an LFM based on a proper correction using high-fidelity information, has been developed to improve the computational efficiency of the probabilistic optimisation process. Since conventional multi-fidelity modelling approaches involve the HFM embracing the same dimension of design spaces as the LFM, the computational cost of such modelling methods is exceptionally high to carry out the probabilistic design of large-scale composite structures. This requires that the HFM call for the same number of high-fidelity FEM simulations as the LFM to build correction response surfaces. It should be noted that even mesh generation for the HFM caused by changes in geometrical random design variables can bring about substantial computational cost. A new multi-fidelity modelling-based probabilistic optimisation method to supervise different design spaces between the HFM and the LFM is developed to mitigate the computational cost in this section. This developed method begins with the critical idea of maximising the use of low-fidelity information while minimising the use of high-fidelity information. The number of high-fidelity design points relies on the dimension of design spaces to be covered. When the HFM has a small number of design variables and the LFM has all design variables, the computational cost to construct the multi-fidelity model can be significantly reduced. In particular, the multi-fidelity modelling method is integrated with multi-level optimisation to manage large-scale composite design problems. As introduced in Chapter 5, the multi-level multi-fidelity modelling method for probabilistic optimisation separately creates two surrogate models of the HFM and LFM with different design variables [51]. This formulation is combined with the RDO of composite structures under the nonlinear post-buckling regime without considering the correlation between the two different fidelity models. The multi-fidelity modelling formulation proposed in this section extends the application area to composite structures under thermo-mechanical loading by considering the correlation between the HFM and the LFM.

6.1.1 *Nonlinear information fusion algorithm*

The multi-fidelity modelling method presented in this chapter employs the nonlinear information fusion algorithm [109], which is also called the nonlinear autoregressive Gaussian process (NARGP), to consider the nonlinear correlation between different fidelities. The NARGP aims to improve the linear autoregressive GP scheme developed by Kennedy and O'Hagan [73]. As discussed in Chapter 3, the autoregressive GP (denoted by AR) is extended by GP regression to construct a probabilistic model that consists of different fidelity models, such as the HFM and the LFM. The generalised form is expressed in equation (6.1):

$$f_{\mathrm{HF}}(X) = z_{\mathrm{LF}}(f_{\mathrm{LF}}(X)) + f_\delta(X) \qquad (6.1)$$

where f_{HF} and f_{LF} are GPs created using the training datasets of the HFM and LFM, respectively. They are usually assigned a zero mean prior $f \sim \mathcal{GP}(f|0, k(x, x'; \theta))$. k is a proper covariance function defined using hyper-parameters θ that produce a covariance matrix, $K_{ij} = k(x_i, x_j; \theta)$, between different input data (x_i, x_j). These hyper-parameters are obtained using the maximum-likelihood estimation [62]. z_{LF} is an unknown function that describes the correlation between the outputs of the HFM and the LFM. f_δ is a GP representing the difference between $z_{\mathrm{LF}}(f_{\mathrm{LF}}(X))$ and $f_{\mathrm{HF}}(X)$.

The NARGP harnesses the functionality of the linear AR without compromising its analytical compliance and straightforward algorithm structure. In particular, $z_{\mathrm{LF}}(f_{\mathrm{LF}}(X))$, the functional composition of two GP priors, is characterised by the so-called deep GP. This does not allow the posterior of f_{HF} to be a Gaussian distribution. To deal with this problem, the NARGP substitutes the GP posterior of low-fidelity predictions, $f_{*\mathrm{LF}}(X)$, with the GP prior, f_{LF} [111]. The principal significance of this nonlinear information fusion algorithm is that z_{LF} and f_{δ} in equation (6.1) incorporate a function, g_{HF}, as described in equation (6.2):

$$f_{\mathrm{HF}}(X) = g_{\mathrm{HF}}(X, f_{*\mathrm{LF}}(X)) \tag{6.2}$$

where $g_{\mathrm{HF}} \sim \mathcal{GP}(f_{\mathrm{HF}}|0, k_{H F_g})((X, f_{*\mathrm{LF}}(X)), (X', f_{*\mathrm{LF}}(X'); \theta_{\mathrm{HF}}))$ is a GP that is characterised by a covariance k_{HF_g} of low-fidelity predictions between different input data (X, X').

The main difference in comparison with the linear AR in equation (6.1) is that f_{δ} is implicitly taken in equation (6.2). Similar to the linear AR, the NARGP under the assumption of noiseless data implies the Markov property, as in equation (6.3), which translates into presuming that given the nearest point of the LFM's posterior, $z_{\mathrm{LF}}(f_{*\mathrm{LF}}(X))$, there is nothing more to learn about $f_{\mathrm{HF}}(X)$ from any other output $z_{\mathrm{LF}}(f_{*\mathrm{LF}}(X'))$ [73]. The assumption of nested training datasets allows the high-fidelity training data, X_{HF}, to be a subset of the low-fidelity training data, X_{LF}:

$$cov\{f_{\mathrm{HF}}(X), z_{\mathrm{LF}}(f_{*\mathrm{LF}}(X'))|z_{\mathrm{LF}}(f_{*\mathrm{LF}}(X))\} = 0, \quad \forall X \neq X' \tag{6.3}$$

It should be noted that the training of g_{HF} using the HFM to conduct the maximum-likelihood estimation becomes more straightforward because the posterior of the LFM, $f_{*\mathrm{LF}}(X_{\mathrm{HF}})$, is a known deterministic quantity. The covariance function of g_{HF} also contributes to the improvement in the NARGP. The function is broken down to consider precise correlations between different input data, as shown in equation (6.4):

$$k_{\mathrm{HF}_g} = k_{\mathrm{HF}_\rho}(X, X'; \theta_{\mathrm{HF}_\rho}) \cdot k_{\mathrm{HF}_f}(f_{*\mathrm{LF}}(X), f_{*\mathrm{LF}}(X'); \theta_{\mathrm{HF}_f})$$
$$+ k_{\mathrm{HF}_\delta}(X, X'; \theta_{\mathrm{HF}_\delta}) \tag{6.4}$$

where k_{HF_ρ}, k_{HF_f}, and k_{HF_δ} are the covariance functions of the scaling factor, the function value from the low-fidelity GP model, and the difference between the two different datasets, respectively. $\theta_{\mathrm{HF}_\rho}$, θ_{HF_f}, and $\theta_{\mathrm{HF}_\delta}$ are the hyper-parameters of each covariance function, which are obtained from the high-fidelity training dataset using the maximum-likelihood estimation, as introduced in Chapter 3.

The covariance function is selected by the square exponential function in equation (3.11). Hence, the NARGP in equation (6.2) enables the LFM's posterior, f_{*LF}, to be projected onto a high-fidelity response, f_{HF}, through a flawless mapping. This allows the multi-fidelity model to capture the nonlinear correlations between the HFM and the LFM. Then, the posterior distributions from the multi-fidelity model are the predictive mean, $\mu_{*MF}(X_*)$, and variance, $\sigma^2_{*MF}(X_*)$. They are calculated using a Monte Carlo simulation that is expressed by

$$p(f_{*HF}(X_*)) := p(f_{HF}(X_*, f_{*LF}(X))|f_{*LF}, X_*, Y_{HF}, X_{HF})$$

$$= \int p(f_{HF}(X_*, f_{*LF}(X))|Y_{HF}, X_{HF}, X_*) p(f_{*LF}(X_*)) \, dX_*$$

$$(6.5)$$

where X_* is a new test point, f_{HF} is the GP of the multi-fidelity model, f_{*LF} is the GP model of the LFM, and Y_{HF} and X_{HF} are the high-fidelity training (input/output) data points. The multi-fidelity model, which provides the predictive mean, $\mu_{*MF}(X_*)$, is incorporated into the proposed multi-fidelity probabilistic optimisation framework.

6.1.2 *Sampling strategy for high- and low-fidelity data*

As shown in Figure 6.1, the formulation aims to maximise the use of the LFM while providing precise corrections using a small number of high-fidelity training data points. Here, the HFM concentrates on the design spaces of a few selected variables at each probabilistic optimisation level. In order to facilitate this purpose, an effective sampling strategy has been employed in this work based on the principles of standard filling sampling strategies such as uniform random sampling or optimal Latin hypercube sampling (OLHS). In general, these space-filling strategies are typically implemented to create a multi-fidelity model due to their nature of offering evenly distributed training data points without gaps or clusters in the whole design space. It is not surprising that the performance of multi-fidelity models relies on how to collect the training data points using an appropriate sampling strategy.

Figure 6.1 illustrates an effective sampling strategy consisting of different sampling degrees to obtain the training data points, including dense and sparse sampling. Each sub-figure in the figure shows a plan view of distribution relying on different design variables to highlight such

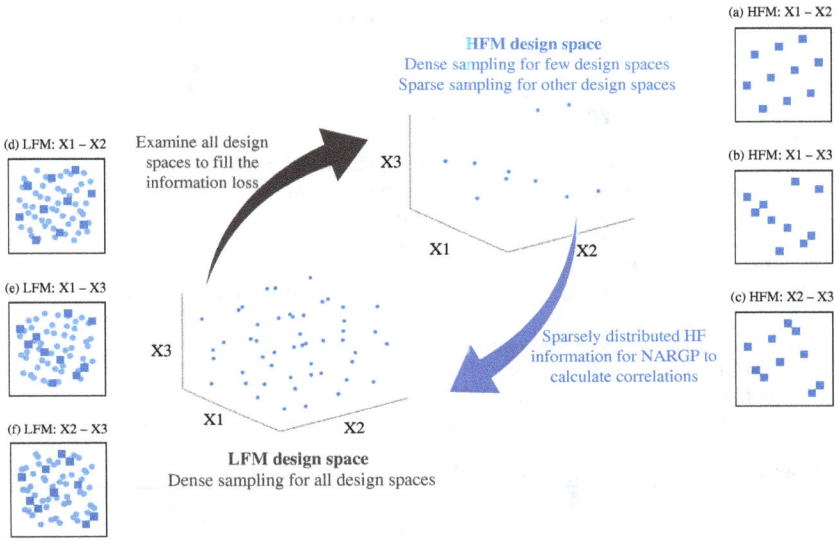

Fig. 6.1. Concept of the proposed multi-fidelity modelling approach.

sampling strategy. Figure 6.1(a) presents the distribution of the dense sampling for the selected design variables in the HFM, H_d in Table 6.1, which refers to collecting evenly enough distributed training data points to embrace the design space of H_d. It should be noted that traditional multi-fidelity modelling methods exploit the training datasets sampled as densely as possible. In comparison, Figures 6.1(b) and 6.1(c) display the distribution of the sparse sampling for other design variables in the HFM, H_s in Table 6.1, which does not thoroughly cover the design space with a biased distribution caused by the insufficient number of training data points. Still, it can provide scarce information within the same sampling size for H_d. Standard regression methods, including GPs or ANNs, commonly require the number of training data points be 10 times more than the dimension of design space [112]. For example, if there are three design variables in a structural optimisation problem in Figure 6.1, the number of training data points, also called the dense sampling in this work, should be at least 30.

This sampling scheme offers a launchpad for the multi-fidelity formulation to reduce the high-fidelity FEM simulations compared with other multi-fidelity modelling methods. Consider the example with three design variables illustrated in Figure 6.1. The HFM focuses on only two design variables ($H_d = [X1, X2]$). In that case, the number of high-fidelity training

Table 6.1.   Sampling strategy for the multi-fidelity formulation.

Iterate
Characterise the design and random variables
Select the design variables in the HFM, H_d
Define other design and random variables in the HFM as H_s
Define all design and random variables in the LFM as L_d
Dense sampling H_d: Around five times the number of design variables in H_d
Sparse sampling H_s: Equal sampling size for H_d
Combine H_d and H_s
Dense sampling L_d: About five times the number of design and random variables in L_d
Create a multi-fidelity model using the NARGP
Evaluate the quality of the model
If satisfied, the model is constructed
If not satisfied, more training data points for the HFM and LFM should be added
Continue until the quality is acceptable

data points is significantly reduced because they cover a two-dimensional design space. Another design variable ($H_s = [X3]$), not selected in H_d, is randomly sampled within the number of training data points for H_d and then added to the high-fidelity training dataset. This allows the high-fidelity training dataset to provide the dense information of the selected design variables, H_d, as well as sparse information of other design variables, H_s, without causing extra computational cost. In the meantime, the LFM exploring the entire design space in Figures 6.1(d)–6.1(f) collects enough information as dense as possible to obtain good quality. The low-fidelity training dataset shares the high-fidelity training data points, as displayed by a blue box in those sub-figures. This allows the NARGP to calculate the correlations using these constructed high- and low-fidelity training datasets based on equation (6.4). Then, it creates the multi-fidelity model using equation (6.2), which can correctly predict output responses, $\mu_{*\mathrm{MF}}(x_*)$, concerning the variation of design variables. It should be highlighted that the number of high-fidelity training data points in this sampling strategy falls well below 30, which the traditional methods require.

6.1.3 *Multi-fidelity formulation and optimisation process*

This section describes how to construct a multi-fidelity model and introduces the multi-fidelity probabilistic optimisation framework. Table 6.2 summarises the workflow that the multi-fidelity modelling formulation

Table 6.2. Multi-fidelity modelling process.

Optimisation level		HFM	LFM
Level I	Design and random variables — Dense sampling	$H_d^{(I)} = [x^{(1)}, \ldots, x^{(m)}]$	$L_d^{(I)} = X = [x^{(1)}, x^{(2)}, \ldots, x^{(n)}]$
	Sparse sampling	$H_s^{(I)} = [x^{(m+1)}, \ldots, x^{(n)}]$	
	Number of design and random variables	n (all variables)	
	Optimal solutions at Level I	$\hat{H}_d^{(I)} = [\hat{x}^{(1)}, \ldots, \hat{x}^{(m)}]$	$\hat{L}_d^{(I)} = [\hat{x}^{(m+1)}, \ldots, \hat{x}^{(n)}]$
	Updated HFM & LFM	$X = [\hat{H}_d^{(I)}, \hat{x}^{(m+1)}, \ldots, \hat{x}^{(n)}]$	
Level II	Design and random variables — Dense sampling	$H_d^{(II)} = [\hat{x}^{(m+1)}, \ldots, \hat{x}^{(l)}]$	$L_d^{(II)} = [\hat{x}^{(m+1)}, \ldots, \hat{x}^{(n)}]$
	Sparse sampling	$H_s^{(II)} = [\hat{x}^{(l+1)}, \ldots, \hat{x}^{(n)}]$	
	Number of design and random variables	$n - m$	
	Optimal solutions at Level II	$\tilde{H}_d^{(II)} = [\tilde{x}^{(m+1)}, \ldots, \tilde{x}^{(l)}]$	$\tilde{L}_d^{(II)} = [\tilde{x}^{(l+1)}, \ldots, \tilde{x}^{(n)}]$
	Updated HFM & LFM	$X = [\hat{H}_d^{(I)}, \tilde{H}_d^{(II)}, \tilde{x}^{(l+1)}, \ldots, \tilde{x}^{(n)}]$	

Find the $H_d^{(III)}$ and $L_d^{(III)}$ using the same method.

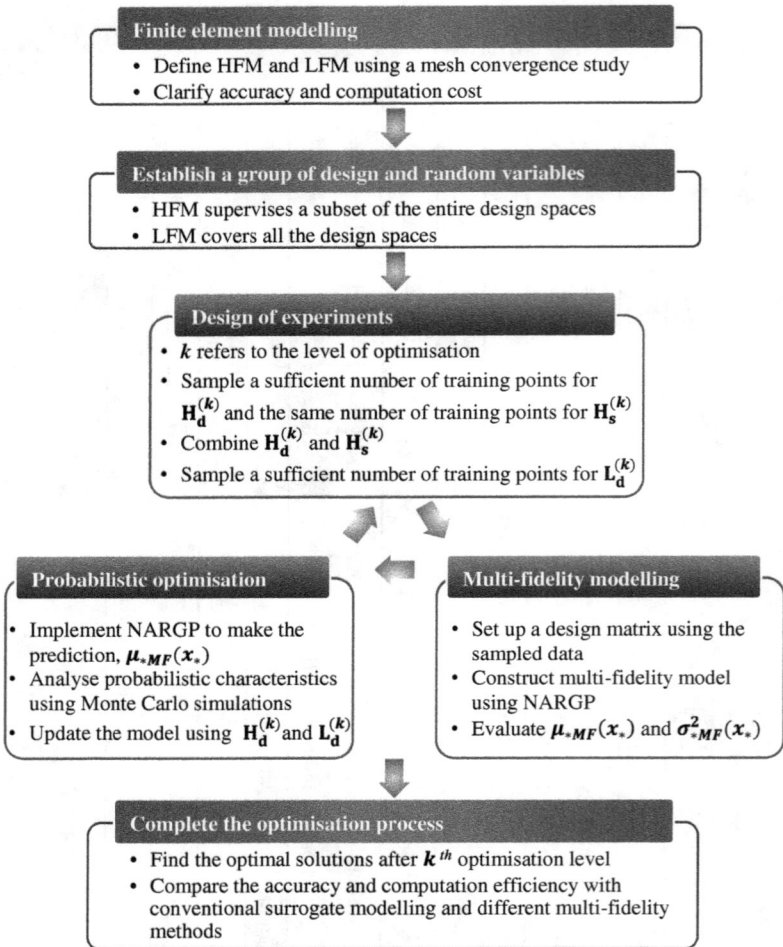

Fig. 6.2.   Multi-fidelity probabilistic optimisation framework.

constructs the multi-fidelity model using the sampling strategy. Then, Figure 6.2 shows the optimisation framework using the constructed multi-fidelity model. The first level of probabilistic optimisation begins with selecting m design variables in the HFM, $H_d^{(I)} = [x^{(1)}, \ldots, x^{(m)}]$, and the training data points for this $H_d^{(I)}$ should be sampled as densely as the NARGP can represent the response surfaces of $H_d^{(I)}$ accurately. The selected design variables, m $(m < n)$, are chosen by the designer from all design variables, n, depending on the problem size. This enables the NARGP to require a smaller number of high-fidelity training data points compared with different multi-fidelity modelling methods considering all design variables in the HFM. Other design variables in the HFM are defined by $H_s^{(I)} = [x^{(m+1)}, \ldots, x^{(n)}]$, and they are not selected in $H_d^{(I)}$. The training data points for $H_s^{(I)}$ are randomly collected within the sampling size for $H_d^{(I)}$ using sparse sampling to preserve the size of the high-fidelity training dataset. Then, these two sampling sets comprising different sampling degrees are combined as a high-fidelity training input dataset for the first level of probabilistic optimisation. At the same time, low-fidelity training data points for the LFM having n design variables, $L_d^{(I)} = [x^{(1)}, x^{(2)}, \ldots, x^{(n)}]$, are also sampled using dense sampling to completely cover the entire design space. Then, two training output datasets corresponding to the high- and low-fidelity training input datasets can be constructed using a proper numerical solver (Abaqus/CAE in this work). The NARGP creates a multi-fidelity model for the first level of probabilistic optimisation using the two high- and low-fidelity training input/output datasets. Figure 6.1 highlights that the multi-fidelity model based on the sampling strategy can provide the correct solutions of $H_d^{(I)}$ during the first level of probabilistic optimisation process since it comprises enough high-fidelity training data points using dense sampling. Simultaneously, the multi-fidelity model examines all design variables' solution spaces using both the sparse high-fidelity training data points for $H_s^{(I)}$ and the dense low-fidelity training data points for $L_d^{(I)}$. When the first level of probabilistic optimisation discovers the optimal solutions, the multi-fidelity model for the second level should be constructed using $H_d^{(II)} = [\hat{x}^{(m+1)}, \ldots, \hat{x}^{(l)}]$, $H_s^{(II)} = [\hat{x}^{(l+1)}, \ldots, \hat{x}^{(n)}]$, and $L_d^{(II)} = [\hat{x}^{(m+1)}, \ldots, \hat{x}^{(l)}]$, followed by updating the HFM and LFM with the optimal solutions of the first optimisation level, as described in Table 6.2 and Figure 6.2.

As shown in Figure 6.1, the cooperation between the HFM and LFM enables the multi-fidelity model to embrace the entire design space without loss of information during the probabilistic optimisation process. Notably, the structure of the NARGP constructs a multi-fidelity model when the training dataset of HFM is a subset of the LFM training dataset. The main contribution of this multi-fidelity formulation is that the HFM focuses on only a small part of the entire design spaces to reduce the sampling size for the high-fidelity training dataset. The HFM also provides sparsely the high-fidelity information of other design variables, $H_s^{(I)}$, while avoiding the requirement of additional high-fidelity FEM simulations. This enables the NARGP to calculate the correlations precisely between the HFM and the LFM based on the sampling strategy to construct the accurate multi-fidelity model. Also, the low-fidelity training dataset using dense sampling complements the lack of information caused by the insufficient high-fidelity information of $H_s^{(I)}$.

Next, the multi-fidelity probabilistic optimisation framework using this proposed multi-fidelity modelling formulation is illustrated in Figure 6.2. Once the NARGP creates the multi-fidelity model using the sampling strategy and the training scheme, the first level of probabilistic optimisation is conducted using the predictions, $\mu_{*\mathrm{MF}}(x_*)$, offered by the constructed multi-fidelity model. The optimal solutions, $\hat{H}_d^{(I)}$, of the selected design variables in the HFM, $H_d^{(I)}$, are found. The design variables, $[x^{(1)}, x^{(2)}, \ldots, x^{(m)}]$, of both the HFM and LFM are updated by the optimal solutions, $X = [\hat{H}_d^{(I)}, \hat{L}_d^{(I)}]$, where $\hat{L}_d^{(I)}$ represents the optimal solutions of other design variables in the LFM, $[x^{(m+1)}, \ldots, x^{(n)}]$. The design variables in $\hat{H}_d^{(I)}$ are fixed and not considered in the second level since their optimal solutions come from the high-fidelity training dataset using dense sampling. Then, different design variables for the second level, $H_d^{(II)} = [\hat{x}^{(m+1)}, \ldots, \hat{x}^{(l)}]$, where $(l < n)$, are selected. The choice of the HFM's design variables should not necessarily follow any particular sequence because the LFM embraces the whole design space at all times. During the second level of probabilistic optimisation, the optimisation algorithm finds the optimal solutions, $\tilde{H}_d^{(II)}$ and $\tilde{L}_d^{(II)}$, of the design variables, $H_d^{(II)}$ and $L_d^{(II)}$, using a new multi-fidelity model constructed for this level. These multi-fidelity modelling and probabilistic optimisation processes are continued until Level k terminates, which means all optimal solutions are found.

The LFM does not require extra FEM simulations to build up a new training dataset for the next level. This can provide additional computational gains compared to the multi-level multi-fidelity method that should establish the low-fidelity training dataset for each level [51]. It is not surprising that even low-fidelity FEM simulations can cause a computational burden when a problem is complex and of large scale. It should be highlighted that the proposed method seeks to harness most of the advantages of multi-fidelity modelling as well as deal with the nonlinear correlation between different fidelity models using nonlinear data fusion GPs. In particular, this method enables the probabilistic design optimisation to broaden its area to the design of composite structures under thermo-mechanical loading.

6.2 Thermo-Mechanical Buckling of Composite Structures

The displacement for a composite structure modelled using three-dimensional shell elements for FEM simulations is based on the first-order deformation plate theory [20], given by equation (6.8):

$$u(x, y, z) = u_0(x, y) + z\phi_y(x, y) \tag{6.6}$$

$$v(x, y, z) = v_0(x, y) + z\phi_x(x, y) \tag{6.7}$$

$$w(x, y, z) = w_0(x, y) \tag{6.8}$$

where u_0, v_0, and w_0 are the displacements at the middle of the plane along each direction, including x, y, and z, respectively. ϕ_x and ϕ_y represent the rotation of the mid-surface in the x and y axes, respectively.

The von Karman strain–displacement relationship to consider geometrical nonlinearity caused by the large deflections of thin plates is described by equation (6.9):

$$E = \begin{bmatrix} \varepsilon_x \\ \varepsilon_y \\ \gamma_{xy} \end{bmatrix} = \begin{bmatrix} \frac{\partial u}{\partial x} \\ \frac{\partial v}{\partial y} \\ \frac{\partial u}{\partial y} + \frac{\partial v}{\partial x} \end{bmatrix} + \frac{1}{2} \begin{bmatrix} \left(\frac{\partial w}{\partial x}\right)^2 \\ \left(\frac{\partial w}{\partial y}\right)^2 \\ 2\frac{\partial w}{\partial x}\frac{\partial w}{\partial y} \end{bmatrix} \tag{6.9}$$

The in-plane strain vector, E, in equation (6.9) is expressed by substituting the equation (6.8):

$$E = \varepsilon^0 - zK = \varepsilon_m + \varepsilon_\theta - zK$$

$$= \begin{bmatrix} \frac{\partial u_0}{\partial x} \\ \frac{\partial v_0}{\partial y} \\ \frac{\partial u_0}{\partial y} + \frac{\partial v_0}{\partial x} \end{bmatrix} + \frac{1}{2} \begin{bmatrix} \left(\frac{\partial w_0}{\partial x}\right)^2 \\ \left(\frac{\partial w_0}{\partial y}\right)^2 \end{bmatrix} - z \begin{bmatrix} \frac{\partial \phi_y}{\partial x} \\ \frac{\partial \phi_x}{\partial y} \\ \frac{\partial \phi_y}{\partial x} + \frac{\partial \phi_x}{\partial y} \end{bmatrix} \tag{6.10}$$

where ε^0, ε_m, ε_θ, and K represent the in-plane strain vector at the mid-plane, the linear in-plane strain vector, the nonlinear in-plane strain vector, and the curvature strain vector, respectively.

In the meantime, transverse shear strains, γ, are expressed as

$$\gamma = \begin{bmatrix} \gamma_{yz} \\ \gamma_{xz} \end{bmatrix} = \begin{bmatrix} \frac{\partial w_0}{\partial y} - \phi_x \\ \frac{\partial w_0}{\partial x} - \phi_y \end{bmatrix} \tag{6.11}$$

In general, equation (6.12) explains the thermo-elastic anisotropic stress–strain relations [1, 20]. It is recognised that thermal stress is not caused by external loads but is a consequence of restrained geometrical thermal distortion:

$$\{\sigma\} = [C]\{[E] - [\alpha]\Delta T\} \tag{6.12}$$

where $[C]$ is a constitutive matrix, $\{\alpha\}$ is the coefficient vector of thermal expansion for a single lamina concerning the in-plane coordinate system, and ΔT is a temperature difference.

The stresses of the kth layer in the laminate layers are calculated using the transformation of coordinates from principal material coordinates, which is represented by equation (6.13):

$$\begin{bmatrix} \sigma_x \\ \sigma_y \\ \tau_{xy} \end{bmatrix} = \begin{bmatrix} Q_{11} & Q_{12} & Q_{16} \\ Q_{12} & Q_{22} & Q_{26} \\ Q_{16} & Q_{26} & Q_{66} \end{bmatrix} \left(\begin{bmatrix} \varepsilon_x \\ \varepsilon_y \\ \gamma_{xy} \end{bmatrix} - \begin{bmatrix} \alpha_x \Delta T \\ \alpha_y \Delta T \\ \alpha_{xy} \Delta T \end{bmatrix} \right) \tag{6.13}$$

where $[Q_{ij}]$ is the transformed stiffness coefficient matrix. α_x, α_y, and α_{xy} are defined as

$$\alpha_x = \alpha_1 \cos^2 \theta + \alpha_2 \sin^2 \theta$$
$$\alpha_y = \alpha_1 \sin^2 \theta + \alpha_2 \cos^2 \theta \tag{6.14}$$
$$\alpha_{xy} = 2(\alpha_1 - \alpha_2) \sin \theta \cos \theta$$

where α_1 and α_2 are thermal expansion coefficients, respectively, and θ is the ply angle of the layer.

The entire force and moment resultants for a N-layered laminate composite structure are defined as

$$\begin{bmatrix} N_x \\ N_y \\ N_{xy} \end{bmatrix} = \int_{-t/2}^{t/2} \begin{bmatrix} \sigma_x \\ \sigma_y \\ \tau_{xy} \end{bmatrix} dz = \sum_{k=1}^{N} \int_{z_{k-1}}^{z_k} \begin{bmatrix} \sigma_x \\ \sigma_y \\ \tau_{xy} \end{bmatrix} dz \tag{6.15}$$

$$\begin{bmatrix} M_x \\ M_y \\ M_{xy} \end{bmatrix} = \int_{-t/2}^{t/2} z \begin{bmatrix} \sigma_x \\ \sigma_y \\ \tau_{xy} \end{bmatrix} dz = \sum_{k=1}^{N} \int_{z_{k-1}}^{z_k} z \begin{bmatrix} \sigma_x \\ \sigma_y \\ \tau_{xy} \end{bmatrix} dz \tag{6.16}$$

where t is the thickness of the kth layer. Note that z_k is the directed distance to the bottom of the kth layer, and z_{k-1} is the directed distance to the top of the kth layer.

When equation (6.13) is substituted in equations (6.15) and (6.16), the force and moment resultants are obtained, which are integrated through the thickness of the composite structure:

$$\begin{Bmatrix} N \\ M \end{Bmatrix} = \begin{bmatrix} A & B \\ B & D \end{bmatrix} \left(\begin{Bmatrix} \varepsilon^0 \\ -\kappa \end{Bmatrix} - \begin{Bmatrix} N_{\mathrm{T}} \\ M_{\mathrm{T}} \end{Bmatrix} \right) \tag{6.17}$$

$$Q = S_\gamma$$

$$(N_{\mathrm{AT}}, M_{\mathrm{AT}}) = \sum_{k=1}^{n} \int_{z_{k-1}}^{z_k} [Q_k][\alpha_k](1, z)\Delta T \, dz \tag{6.18}$$

where the laminate stiffness is defined as $[A], [B], [D]$ are defined by integrations over the laminate thickness:

$$[A], [B], [D] = \sum_{k=1}^{n} \int_{z_{k-1}}^{z_k} [Q_{ij}](k, z, z^2) \, dz \quad (i, j = 1, 2, 6) \tag{6.19}$$

$$[S] = \sum_{k=1}^{n} \kappa_k \int_{z_{k-1}}^{z_k} [Q_{ij}] \, dz \quad (i, j = 4, 5) \tag{6.20}$$

κ_p is a shear correction factor, and $[N_{\Delta T}]$ and $[M_{\Delta T}]$ are the thermal forces and thermal moment induced by the temperature changes ΔT, respectively.

The principle of virtual work derives the governing equation for thermo-mechanical buckling:

$$\delta W = \delta W_{\mathrm{int}} - \mathcal{E} W_{\mathrm{ext}} = 0 \tag{6.21}$$

where δW_{int} is the internal virtual work that consists of $\delta W_{\mathrm{int}}^{(1)}$ and $\delta W_{\mathrm{int}}^{(2)}$ done by the linear and thermal geometric stiffness components.

$$\delta W_{\mathrm{int}}^{(1)} = \int_A \left(\delta \varepsilon^T N + \delta \kappa^T M \right) dA$$
$$\delta W_{\mathrm{int}}^{(2)} = \int_A \left(\delta \theta^T Q \right) dA \tag{6.22}$$

where d is the displacement vector and $[K]$ is the linear stiffness matrix.

Thermal geometric stiffness matrix, $[K_{\Delta T}]$, is offered by the work done by a constant thermal force that leads to a small lateral deflection. This internal work is done by the thermal forces, including $N_{\Delta Tx}$, $N_{\Delta Ty}$, and

$N_{\Delta Txy}$, caused by the temperature change. The change in strain energy due to the thermal forces is written as equation (6.23):

$$
\begin{aligned}
\delta W_{\text{int}}^{(2)} &= -\int_A \left[\left[\delta\varepsilon^T \ N_{\Delta T}\right]\right] dA \\
&= -\int_A \left[\left[N_{\Delta Tx} \ N_{\Delta Ty} \ N_{\Delta Txy}\right] \delta \begin{bmatrix} \frac{\partial w}{\partial x} \\ \frac{\partial w}{\partial y} \end{bmatrix}\right] dA \\
&= -\int_A \delta w^T \begin{bmatrix} N_{\Delta Tx} & N_{\Delta Txy} \\ N_{\Delta Txy} & N_{\Delta Ty} \end{bmatrix} \begin{bmatrix} \frac{\partial w}{\partial x} \\ \frac{\partial w}{\partial y} \end{bmatrix} dxdy \\
&= -\{\delta d\}^T [K_{\Delta T}]\{d\}
\end{aligned}
\tag{6.23}
$$

Finally, the equation of motion of the composite structure and the eigenproblem for the thermo-mechanical buckling analysis are obtained, which can be represented as

$$
([K] - \lambda[K_{\Delta T}])\{d\} = \{0\}
\tag{6.24}
$$

$$
([K] - \lambda[K_{\Delta T}])\{d\} = \{0\}
\tag{6.25}
$$

where λ and $\{d\}$ are the critical temperature change and buckling mode shape, respectively.

6.3 Numerical Example

The proposed multi-fidelity probabilistic optimisation framework was demonstrated using the RBDO of a mono-stringer stiffened composite panel under thermomechanical loading. This demonstration shows the potential of the developed multi-fidelity optimisation method to be utilised for the probabilistic design of large-scale composite structures under both mechanical and thermal loading. Notably, the computational efficiency of the presented method was highlighted by comparison with traditional multi-fidelity models and high-fidelity surrogate models.

6.3.1 *Composite structures under thermo-mechanical loading*

The details of the mono-stringer stiffened composite structure are illustrated in Figure 6.3, which is identical to the composite structures considered in previous chapters. This structure is clamped at both ends, but

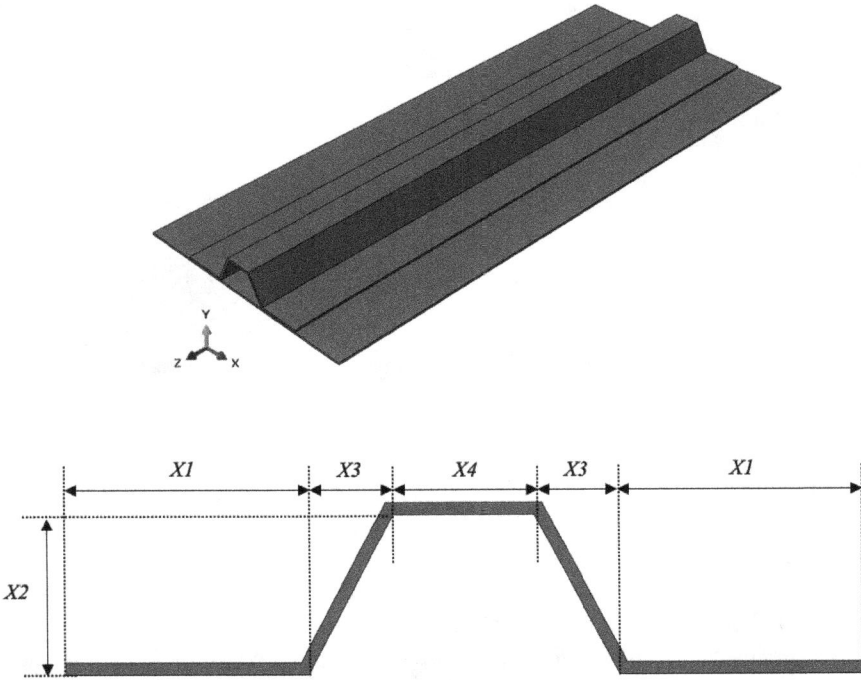

Fig. 6.3. Mono-stringer stiffened composite panel.

the left-hand end is free to move in the longitudinal direction (z-direction in the figure), which is the applied loading direction. A pure compression load for mechanical shortening is applied by increasing the uniform displacement at the left-hand end. The material properties are the same as in Table 4.1. The thermal expansion coefficients of the structure are shown in Table 6.3. It should be noted that only the stringer geometry is to be optimised, while the optimisation process considers the uncertainties in the geometric and mechanical properties of both the stringer and the skin. There are no constraints on the two longitudinal edges of the skin. Perfect bonding is assumed between the stiffener and skin to consider their interaction.

The thermo-mechanical buckling analysis depending on different mechanical shortening lengths was conducted to see how significant the critical temperature changes are. Figure 6 4 displays the thermo-mechanical buckling results for a mono-stiffened stringer panel with the mean geometry

Table 6.3.   Thermal expansion coefficient.

Parameter	Notation	Value
Longitudinal thermal coefficient	α_1	1.7×10^{-6}
Transverse thermal coefficient	α_2	-1×10^{-6}

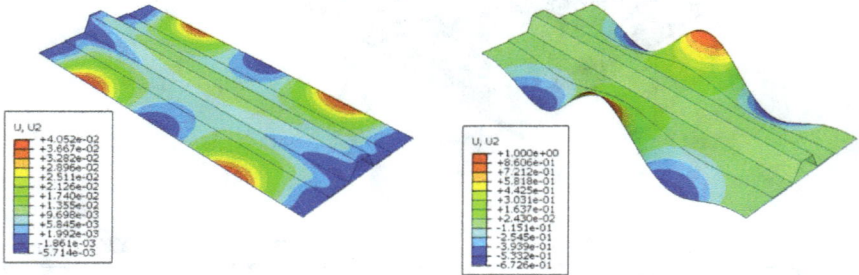

Fig. 6.4.   Out-of-plane displacements of mechanical shortening (left) and temperature rise (right).

values at the design space. As seen in the figure, the composite structure experiences mechanical shortening caused by pure compression, and then a thermal buckling analysis is conducted to find the critical temperature change. Table 6.4 shows that the thermo-mechanical buckling occurs in the vicinity of 95°C, which is the normal operating temperature range of regional aircraft when the shortening length is 0.3 mm ($\Delta L/L = 0.05\%$). In this work, the thermo-mechanical buckling temperature refers to the maximum critical temperature change followed by the shortening, $\Delta L/L = 0.05\%$. The class of fidelity was decided by the level of FEM discretisation. Figure 6.5 illustrates the mesh grid of both the HFM and LFM; the element size was defined as 4.0 and 12.0 mm, respectively, through a mesh convergence study for the thermo-mechanical buckling temperature. It should be highlighted that the LFM shows around 15% error while demonstrating a computational cost of only about 30% compared to the HFM. The FEM models are composed of four-node shell elements (S4R).

6.3.2 *Multi-fidelity modelling*

A multi-fidelity model using the proposed formulation is constructed to carry out the probabilistic optimisation of composite structures under

Table 6.4. Critical temperature changes depending on mechanical shortening.

Mechanical shortening (mm)	0	0.1	0.2	0.3
Critical temperature changes (°C)	407	303	200	95

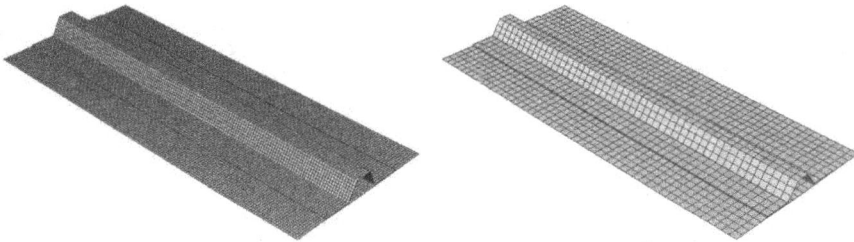

Fig. 6.5. 4 mm mesh size for HFM (left) and 12 mm mesh size for LFM (right).

thermo-mechanical loading. The input parameters of the multi-fidelity model are the mono-stiffened stringer geometry ($X1$, $X2$, $X3$, and $X4$) in Figure 6.3 and the mechanical properties of the composite structure (E_{11}, $E_{22} = E_{33}$, G_{23}, $G_{12} = G_{13}$, α_{11}, and α_{22}) in Tables 4.1 and 6.3. These input parameters are defined as the design and random variables that are used for optimisation and reliability analysis, respectively. In this work, the design variables are geometric parameters, whereas the random variables are both geometric and mechanical properties. Outputs are the critical temperature change and the mass of the composite structure. This multi-fidelity model having ten input parameters and two output parameters is constructed using both the HFM and the LFM covering different design spaces. As mentioned before, the NARGP quantifies nonlinear correlations between different fidelities to create an accurate multi-fidelity model.

First, the training datasets for the HFM and LFM should be sampled using the OLHS technique, which is a sampling technique to collect training data in the given design space as evenly distributed as possible. Table 6.5 shows the 10 input parameters, each having a design range that consists of four geometric parameters and six mechanical properties. The geometric parameters primarily influence the output of the composite structure. At the same time, they have uncertainties associated with design and manufacturing that could considerably affect the structure's performance. Apart from these geometric parameters, the uncertainties of the mechanical

Table 6.5.   Design and random variables.

Design of experiment input data	Value
Stringer foot (mm)	$34.4 < X1 < 51.6$
Stringer height (mm)	$24.0 < X2 < 36.0$
Distance between top and bottom (mm)	$12.0 < X3 < 18.0$
Stringer top (mm)	$20.0 < X4 < 30.0$
E_{11} (GPa)	$111 < E_{11} < 167$
$E_{22} = E_{33}$ (GPa)	$6.5 < E_{22} < 9.8$
$G_{12} = G_{13}$ (GPa)	$3.8 < G_{12} < 5.8$
G_{23} (GPa)	$2.5 < G_{23} < 3.7$
α_{11}	$1.4E^{-6} < \alpha_{11} < 2.0E^{-6}$
α_{22}	$-1.2E^{-6} < \alpha_{22} < -0.8E^{-6}$

Table 6.6.   Details of the multi-fidelity models.

Level	Scatter degree of training data	HFM Design and random variables	Number of FEM simulations	LFM Design and random variables	Number of FEM simulations
I	Dense	$X1$, $X2$	10	$X1$, $X2$, $X3$, $X4$, E_{11}, $E_{22} = E_{33}$, G_{23}, $G_{12} = G_{13}$, α_{11}, α_{22}	60
	Sparse	$X3$, $X4$, E_{11}, $E_{22} = E_{33}$, G_{23}, $G_{12} = G_{13}$, α_{11}, α_{22}		—	
II	Dense	$X3$, $X4$	10	$X3$, $X4$, E_{11}, $E_{22} = E_{33}$, G_{23}, $G_{12} = G_{13}$, α_{11}, α_{22}	10
	Sparse	E_{11}, $E_{22} = E_{33}$, G_{23}, $G_{12} = G_{13}$, α_{11}, α_{22}		—	

properties should be considered as well. Table 6.6 clearly shows how the multi-fidelity model is constructed using two different fidelity models that explore and supervise different design spaces at each level. The number of FEM simulations to construct the multi-fidelity model is also shown in the table. As mentioned in the previous section, the HFM covers only a part of the whole design space as dense as the sampled high-fidelity dataset can precisely examine the response surfaces of the selected design variables, $X1$ and $X2$, in this work. When the HFM at the first level focuses on the two design variables, 10 training data points are enough to create the metamodel concerning these design variables accurately. The same number of training data of other design variables should also be sampled sparsely and added to the high-fidelity training dataset. Although the number of 10 training data points is not sufficient to set up other design variables' metamodel, it allows the NARGP to use the high-fidelity information as much as possible in the given training dataset without the need for additional high-fidelity FEM simulations. In contrast, the LFM, because it supervises all design variables, requires 60 training data points to supplement the lack of other design variables' information that is not included in the HFM. This training dataset embraces the response surfaces of all design variables as solidly as possible. Once the NARGP constructs the multi-fidelity model, error analysis should be conducted using a test dataset that uses points not included in the training scheme. The created multi-fidelity model at each level predicts the critical temperature change and the mass with less than 1.5% error in the given design spaces.

When the multi-fidelity model completes the first level of optimisation, this multi-fidelity model provides the optimal solutions of the two selected design variables in the HFM as well as the corresponding values of other design variables in the LFM. These optimal solutions update the HFM to choose different design variables that are not considered at the previous level. The LFM is also updated by those optimal solutions that enable a smaller number of design variables than the first level. The corresponding values obtained by the LFM update other design variables that are not considered in HFM. After updating the two models, the HFM chooses new design variables, $X3$ and $X4$, as shown in Table 6.6. Then, 10 new training data points that can correctly establish the chosen design variables' response surfaces are sampled using OLHS as densely as possible. Simultaneously, the same number of training data points of other design variables are also collected sparsely to build up an efficient high-fidelity training dataset. It is not surprising that the LFM at the first level is still precise enough to carry out the next-level optimisation since it is made up of 60 training points. This means the LFM requires only 10 additional low-fidelity FEM simulations that are identical to the training data of HFM for this level since the high-fidelity training dataset should be a subset of the low-fidelity training dataset. Figure 6.6 highlights how the HFM and LFM cooperate using different sampling levels in different design spaces when $X1$ and $X2$ are selected as the HFM's design variables. The high-fidelity training dataset is evenly distributed using the small number of training points in the design space of $X1$ and $X2$. However, it does not seem that the rest of the design spaces, $X3$ and $X4$, are scattered uniformly in the high-fidelity design spaces. The low-fidelity training dataset carefully covers the whole design space without information loss, dissimilar to the high-fidelity design spaces. It should be highlighted that this multi-fidelity scheme enables the size of the high-fidelity training dataset to decrease while embracing the entire design space with sparsely distributed high-fidelity information as well as dense low-fidelity information. It should be noted that conventional surrogate modelling approaches require hundreds of training points to consider all design and random variables. Some multi-fidelity modelling methods also demand a significant number of high-fidelity training data points because the HFM encompasses the same design spaces as the LFM. The proposed multi-fidelity modelling methodology provides significant computational time savings compared with other multi-fidelity methods and enables probabilistic optimisation to broaden its application area to large-scale composite structures under thermo-mechanical loading.

6.3.3 *Multi-fidelity probabilistic design optimisation*

RBDO, a type of probabilistic optimisation, was conducted to demonstrate the efficiency and accuracy of the multi-fidelity model constructed using the proposed formulation. In this example, geometric nonlinearities were considered, while the material properties were presumed to be in the linear elastic region.

6.3.3.1 *Problem definition*

Table 6.7 provides a description of the RBDO problem of the mono-stiffened stringer composite structure under thermomechanical loading.

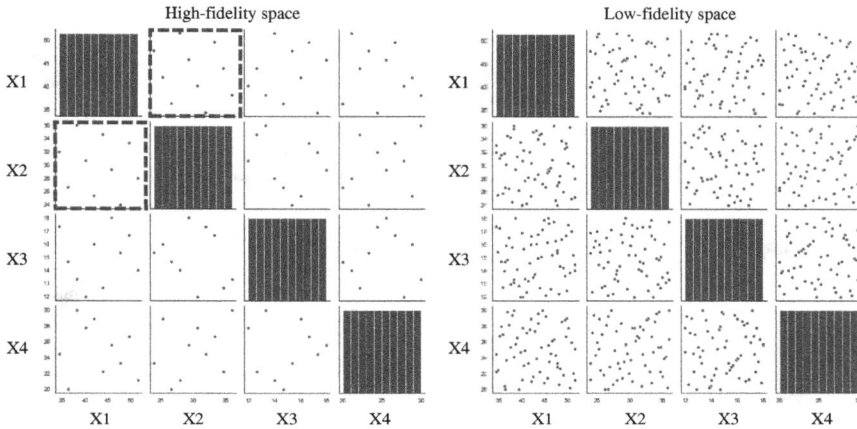

Fig. 6.6.   Training data distribution between HFM and LFM.

There are four constraints and three objectives. Since this proposed multi-fidelity optimisation approach aims to cooperate with the HFM and the LFM exploring different design spaces, each fidelity model has its own constraints and objectives in the optimisation process. They carry typical mass constraints but different constraints regarding minimum critical temperature changes based on the difference between the HFM and the LFM. The reliability index was targeted by the associated value with the probability of failure of 0.135%. The objective functions were to maximise the critical temperature changes and minimise the mass of the composite structure. It should be highlighted that the composite structure should be under thermal conditions followed by a specific mechanical shortening of 0.05% regarding the longitudinal direction to include the effect of residual stress.

The discretisation level of the FEM models for the HFM and the LFM were defined through a mesh convergence study. It should be noted that the LFM shows 15% solution error compared to the HFM while offering a 70% reduction in computational cost. The design spaces of geometric parameters and material properties are the same as the range of design of experiments in the previous section. As shown in Figure 6.7, the optimisation was conducted using NSGA-II [45], a multi-objective exploratory technique. This optimisation method is suitable for highly nonlinear design spaces as well as discontinuous design spaces. The method

Table 6.7.   Reliability-based design optimisation: problem definition.

Description		Value
Multi-fidelity model	HFM	*mesh size*: 4.0 mm
	LFM	*mesh size*: 12.0 mm
Optimisation method	NSGA-II	*generation*: 12
		population: 10
Analysis type	Monte Carlo simulations	
Design uncertainty	Mean	Standard deviations
	$X1$	$0.001 \times X1$
	$X2$	$0.001 \times X2$
	$X3$	$0.001 \times X3$
	$X4$	$0.001 \times X4$
	E_{11}	$0.05 \times E_{11}$
	E_{22}	$0.05 \times E_{22}$
	G_{12}	$0.05 \times G_{12}$
	G_{23}	$0.05 \times G_{23}$
	α_{11}	$0.05 \times \alpha_{11}$
	α_{22}	$0.05 \times \alpha_{22}$
Constraints under 0.05%	Mass	$m \leq 1$ kg
mechanical shortening	HFM ΔT_{cr}	$\Delta T_{cr,HFM} \geq 50$
	LFM ΔT_{cr}	$\Delta T_{cr,LFM} \geq 57.5$
	Reliability index	$\beta = 3$
Constraints under 0.05%	Mean mass	*minimise*
mechanical shortening	Mean HFM ΔT_{cr}	*maximise*
	Mean LFM ΔT_{cr}	*maximise*

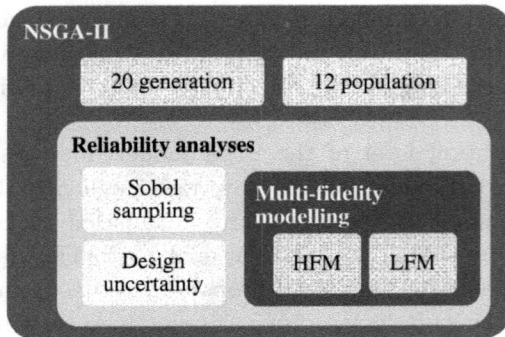

Fig. 6.7.   Multi-fidelity RBDO framework.

follows the standard genetic operation of mutation and crossover. Still, the selection process is based on different mechanisms to construct a Pareto set with the best combination of objective values. Generation and population numbers were determined as 12 and 20, respectively, to find the optimal solutions correctly. These values were obtained using the convergence check of the Pareto front depending on different combinations of generation and population. In particular, the reliability check to consider the design uncertainties of random design variables is essential in this optimisation process. The uncertainties associated with the geometric parameters are commonly assumed to be a 0.1% coefficient of variation concerning its mean values as manufacturing tolerance. The uncertainties of mechanical properties are considered by a random normal distribution with a 5% coefficient of variation [16]. All input parameters are presumed to have a truncated Gaussian distribution at three standard deviations. The reliability assessment should be conducted at all populations of each generation. Monte Carlo simulation computes the statistical characteristics of the objective functions, which are induced by the uncertainties of random design variables. The Sobol sampling method was incorporated into the Monte Carlo simulation to obtain a more homogeneous sampling distribution as well as more robust statistical estimations than other sampling methods [58]. It should be noted that the maximum allowable number of multi-fidelity simulations to check the probability of failure was set to 2,000. The convergence tolerance check was carried out at every 25 sampling points to improve the computational efficiency. The Monte Carlo simulations were halted when both the mean and the standard deviations satisfy a 0.1% difference with these associated values at the previous convergence test. Hence, the maximum simulation number using the developed multi-fidelity model was 480,000 for the two levels in this RBDO process.

6.3.3.2 *Results*

Table 6.8 presents the RBDO results obtained using the proposed multi-fidelity formulation. This table demonstrates how the multi-fidelity model enables cooperation between two different fidelity models and improves its accuracy at the end of each level. In the table, the HFM has two different sampling degrees of training data, while the LFM is sampled densely as it can produce the response surfaces for all random design variables. At the first level, the HFM carries two design variables, $X1$ and $X2$, while

Table 6.8.   RBDO results.

Level	Scatter degree of training data	Design and random variables used for the HFM and LFM		Result			Optimal design values to update the multi-fidelity model
		HFM	LFM	Critical temperature change		Mass	
				HFM	LFM		
I	Dense	$X1$, $X2$	$X1$, $X2$, $X3$, $X4$, E_{11}, $E_{22} = E_{33}$, G_{23}, $G_{12} = G_{13}$, α_{11}, α_{22}	87	95	861	$X1$: 34.67, $X2$: 24.10, $X3$: 17.82, $X4$: 27.27
	Sparse	$X3$, $X4$, E_{11}, $E_{22} = E_{33}$, G_{23}, $G_{12} = G_{13}$, α_{11}, α_{22}	—				
II	Dense	$X3$, $X4$,	$X3$, $X4$, E_{11}, $E_{22} = E_{33}$, $G_{12} = G_{13}$, α_{11}, α_{22}	98	110	867	$X1$: 34.67, $X2$: 24.10, $X3$: 18.00, $X4$: 30.00
	Sparse	E_{11}, $E_{22} = E_{33}$, G_{23}, $G_{12} = G_{13}$, α_{11}, α_{22}	—				

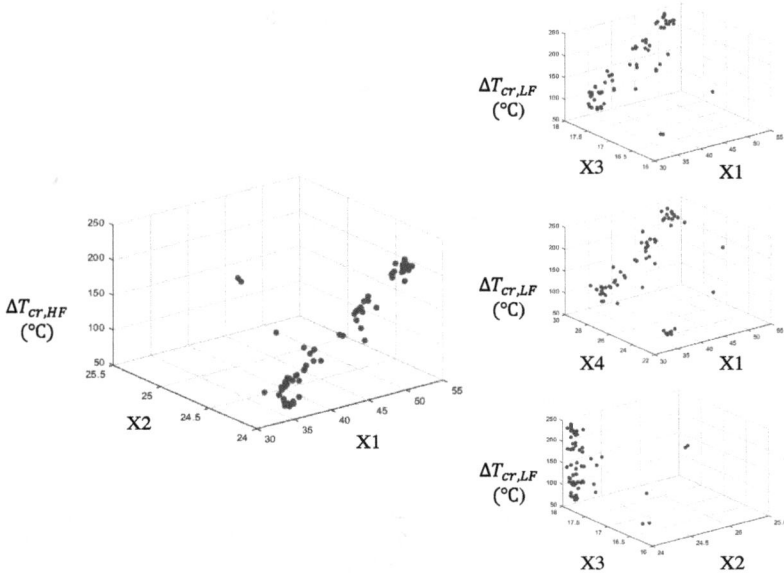

Fig. 6.8.   Multi-fidelity modelling-based RBDO at Level I.

the LFM has all the random design variables. At the same time, the HFM takes the other two design variables, $X3$ and $X4$, and six random variables of material properties as a form of sparsely scattered training dataset that allows the LFM to quantify correlations among all variables. Both the HFM and LFM share the selected design variables of $X1$ and $X2$ during the first level of the optimisation. As shown in Figure 6.8, this enables the multi-fidelity model to scrutinise the solution spaces of other design variables corresponding to those of the selected design variables. When the first level finds the Pareto front satisfied with the objectives and constraints, the optimal solutions should be chosen by the designer's decision. In this work, the optimal solutions were chosen by the allowable temperature range associated with the regional aircraft operation, known to vary from $-45°C$ to $85°C$. The chosen optimal solutions at the first level update the multi-fidelity model before the next level. The optimal solutions of $X1$ and $X2$ are good enough to be fixed at the next level because they are obtained by dense high-fidelity information. It is not surprising that the corresponding optimal solutions of the LFM at the first level can update the other design variables, $X3$ and $X4$, that are not included in the HFM. Hence, this update enables the HFM to take different design variables that are collected as densely

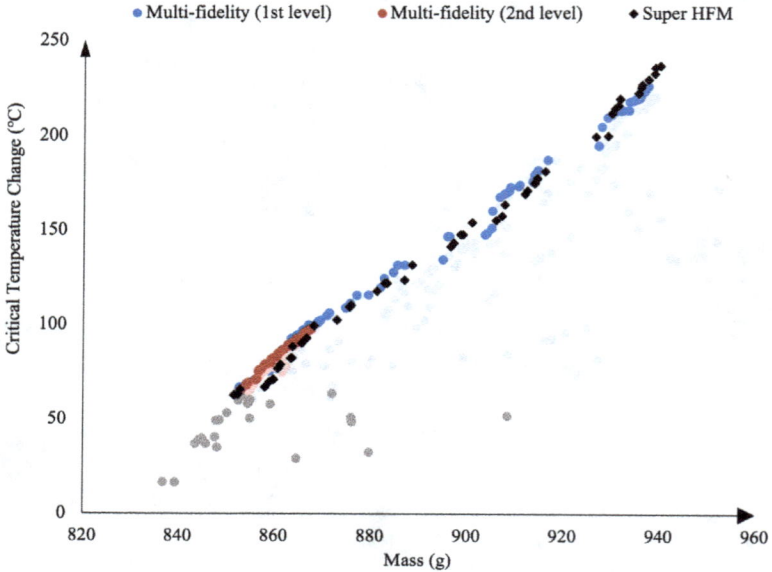

Fig. 6.9. RBDO results comparison by Pareto front for multi-fidelity model vs. super HFM.

Note: Bold points are the Pareto front, while fainter points are feasible solutions.

as possible. Similarly as the first level, this updated HFM carries different design variables, $X3$ and $X4$, while the information of six random variables is also included sparsely. The LFM is updated by the optimal solutions of $X1$ and $X2$, while the HFM takes all random design variables except for the fixed $X1$ and $X2$. When this level is finished, the final solution is obtained because there are two levels in this example.

Figure 6.9 illustrates the Pareto fronts of the optimisation results at each level. The Pareto front of the first level is associated with the selected design variables that the HFM mainly supervises for the optimisation process, while the LFM compensates for the lack of high-fidelity information. The second level's Pareto front has a small range since the optimal solutions in the first level's Pareto front update the multi-fidelity model. It should be highlighted that the second level examines the solution spaces thoroughly based on the chosen solution of the first level and discovers more reliable and better solutions compared to those of the first level. As can be seen in the figure, the Pareto front points at the second level rise in the upward direction and fill the gap in the solution space left

Table 6.9.   Design value and finite element analysis check.

Model	Design variables (mm)				Mass (g)	Approximated model $\Delta T_{cr}(°C)$		FEM model ΔT_{cr} (°C)
	X1	X2	X3	X4		μ	σ	
MF model	34.67	24.10	18.00	30.00	867	98.1	5.2	98.7
Super HFM	35.35	24.01	17.97	29.08	868	99.5	12.1	98.7
HFM40	37.25	24.01	17.97	23.90	865	87.4	9.3	85.1

from the first level. The figure also includes the Pareto front found from a conventional surrogate modelling approach, which is called a super HFM in this work, to highlight how accurate the proposed multi-fidelity method is. The super HFM consists of 150 high-fidelity FEM simulations to correctly resemble 10 random design variables' response surfaces. This figure shows that the Pareto front of the super HFM is nearly identical to that of the multi-fidelity model.

Table 6.9 displays the mean value and standard deviation of the chosen optimal solutions so that the results of reliability assessments are compared between the multi-fidelity model and the super HFM. It should be noted that the optimal solution of the super HFM is selected as having the same mass as the chosen optimal solution of the multi-fidelity model. The multi-fidelity model has almost the same mean value and acceptable standard deviation as those of super HFM. It is also significant to assess the accuracy of the multi-fidelity model as the solution of the HFM created by the equivalent number of high-fidelity FEM simulations. There could be no advantages to the proposed method unless the multi-fidelity model constructed at the same computational cost is more accurate. Table 6.6 shows that the multi-fidelity model comprises 20 HFMs and 70 LFMs, which equate to 40 high-fidelity FEM simulations. Figure 6.10 features the accuracy among three different models: the multi-fidelity model, the super HFM, and the equivalent HFM (HFM40). In this figure, the multi-fidelity model's mean value is much closer to that of the super HFM than that of HFM40 while having a standard deviation similar to the super HFM. This means that the reliability assessments of the proposed multi-fidelity method are more accurate than those obtained from an equivalent number of high-fidelity FEM simulations in terms of computation time. As the NARGP is incorporated into the multi-fidelity formulation, the constructed multi-fidelity model has a black-box structure between input and output

168   *Probabilistic Optimisation of Composite Structures*

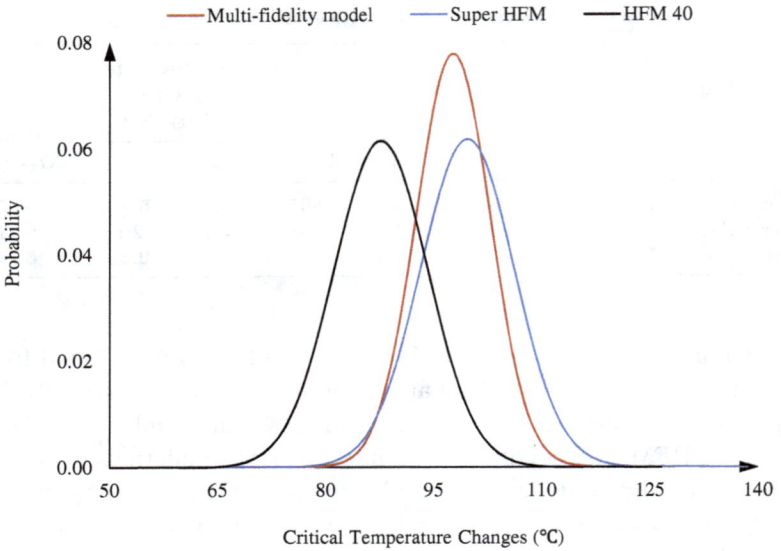

Fig. 6.10.   Statistical comparison between three different models.

parameters. The chosen optimal solutions should be checked to show if the solutions make sense using proper techniques, such as a FEM solver or an experiment. Table 6.9 shows the critical temperature change using the multi-fidelity model is nearly identical to that of the FEM solver. It should be remarked that the accuracy of the multi-fidelity model is greater than that of the super HFM at this design point.

The key benefit of multi-fidelity modelling methods is that they can reduce the significant computational cost caused by the consideration of uncertainties during the probabilistic optimisation process. Figure 6.11 represents the computational time savings of the proposed multi-fidelity method compared to different probabilistic optimisation methods. All computational costs in this figure are normalised by the total computational cost of the super HFM so that the computational gains are emphasised in a practical way [57]. The proposed multi-fidelity model is constructed using both 20 HFM training points and 70 LFM training points. The equivalent computational cost equates to about 40 high-fidelity FEM simulations because the computational cost of the LFM is 70% more economical than the HFM. In contrast, the super HFM requires 150 high-fidelity FEM simulations. It is not surprising that the proposed multi-fidelity optimisation method provides around 75% of computation time savings.

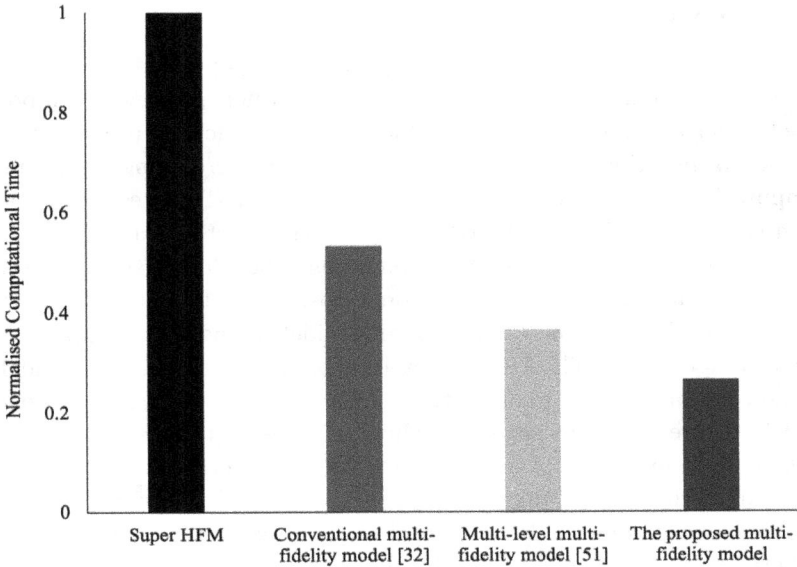

Fig. 6.11.   Computational efficiency comparison between four different methods.

However, it is more significant to show how much computation time savings are obtained by the use of this new method compared with different multi-fidelity methods. Traditional multi-fidelity methods [29] construct a multi-fidelity model using both 60 HFM training points and 60 LFM training points because two different fidelity models have the same number of design variables. The equivalent high-fidelity FEM simulation number for this model is around 80, which is twice as computationally expensive as the proposed multi-fidelity method. This proposed multi-fidelity method is still more efficient than the multi-level multi-fidelity modelling approach developed earlier [51]. The multi-level multi-fidelity approach uses both 20 HFM training points and 110 LFM training points, which equates to around 55 high-fidelity FEM simulations; hence, the proposed multi-fidelity method enables 30% of computation time savings. Although the LFM is generally computationally cheap, its computational cost can cause a severe burden when it comes to large-scale problems. Therefore, the proposed novel multi-fidelity method allows for considerably significant computational benefits to broaden the application range of the multi-fidelity method to the probabilistic design of large-scale composite structures.

6.4 Summary

In this work, a novel multi-fidelity probabilistic optimisation approach was proposed and demonstrated with a mono-stiffened stringer composite panel under thermo-mechanical loading. This approach enables the probabilistic optimisation of composite structures while providing considerable computation time savings and reliable solutions. This research work's main contribution is that the HFM supervises a part of the entire design space as densely as possible while examining other design spaces sparsely while not causing extra computational cost. Simultaneously, the LFM takes all design variables for the multi-fidelity model to explore the whole design space. The NARGP is incorporated into this multi-fidelity method because it can quantify the correlations between different fidelity models to predict more accurate solutions. This method also cooperates with a multi-level optimisation framework to be utilised for large-scale problems with a vast number of design variables. The proposed method was demonstrated by the RBDO problem of the composite structure under a thermal environment followed by mechanical shortening. The optimal solutions and their reliability assessments obtained by the proposed multi-fidelity method were nearly identical to those of the super HFM, the conventional surrogate modelling approach. They were also more accurate than a surrogate model that used the equivalent number of high-fidelity FEM simulations to the multi-fidelity model in terms of computation time. Compared with a FEM solver, the multi-fidelity model predicts the critical temperature changes more precisely than those of the super HFM for this example. As well as improvements in accuracy, this proposed method provides significant improvements to computational efficiency. This new method's computational efficiency was highlighted by comparing it with a conventional surrogate method, traditional multi-fidelity method, and multi-level multi-fidelity method. It should be noted that the proposed method offers remarkable computation time savings. In particular, this method is 30–50% more computationally efficient than other multi-fidelity methods. The developed multi-fidelity probabilistic optimisation method is a new optimisation approach that enables the HFM and the LFM to supervise different design spaces while cooperating with each other during the optimisation process. This new multi-fidelity method enables the probabilistic optimisation of complex and large-scale problems, such as for composite structures under thermo-mechanical loading, to be conducted in a significantly more efficient and accurate manner.

Chapter 7

Conclusion

In this chapter, general conclusions and potential suggestions for both short- and long-term future work are discussed.

As introduced in Chapter 1, the main objective of the work presented in this monograph was to develop novel multi-fidelity modelling-based probabilistic design optimisation methods for composite structures, including reliability-based design optimisation (RBDO) and robust design optimisation (RDO). In order to achieve this objective, new multi-fidelity modelling formulations were formulated to specifically reduce the computational costs associated with the high-fidelity finite element analysis. The newly developed formulations bridged the gap between high-fidelity models (HFMs) and low-fidelity models (LFMs) using machine learning techniques, including artificial neural networks (ANNs) and nonlinear autoregressive Gaussian processes (NARGPs). Furthermore, a multi-level optimisation approach was developed, and a new sampling strategy was integrated into the probabilistic design for the first time. The developed multi-fidelity probabilistic optimisation methods enable HFMs and LFMs to have a different number of design variables during the optimisation process, thereby offering more computational benefits than traditional surrogate methods and existing multi-fidelity methods. Several engineering examples using aircraft mono-stringer stiffened composite panels under mechanical and thermal loads demonstrated the accuracy and computational efficiency of the developed multi-fidelity probabilistic optimisation methods. The results indicated that the developed methods significantly reduce the required computational time, allowing for more design variables to be considered early in the design stage of large-scale and complex aircraft composite structures. The results discussed in Chapters 4–6 show that the primary research aim has been accomplished.

In Chapter 4, a multi-fidelity modelling-based RBDO framework was developed for composite structures for the first time. This framework proved to be capable of providing acceptable accuracy and substantial computational efficiency when compared with the conventional RBDO optimisation method based on the commonly used high-fidelity surrogate model. The new multi-fidelity framework also assured the capability to carry out different reliability analyses for estimating the probability of failure. These were demonstrated in two engineering examples at the end of the chapter, highlighting the accuracy and efficiency of the developed framework.

In Chapter 5, a new multi-fidelity modelling-based RDO framework was developed to cover different design spaces between the HFM and the LFM. Two engineering examples, featuring a mono-stiffened stringer composite panel subjected to the nonlinear post-buckling regime, were examined. The developed framework integrated with multi-level optimisation produced additional computational gains compared to traditional multi-fidelity methods while maintaining acceptable accuracy.

In Chapter 6, a novel multi-fidelity modelling-based probabilistic optimisation framework was developed for considering nonlinear correlations between the HFM and the LFM using Gaussian processes (GPs). An engineering example of the same composite panel under thermo-mechanical loading was investigated. This new framework, employing different sampling degrees for the HFM, proved to be more computationally efficient than other different multi-fidelity methods. The accuracy of this framework was further improved through the use of nonlinear fusion GPs.

As mentioned in Chapter 1, the primary objectives of the work presented in this monograph can be broken down into five sub-objectives. The conclusions for each of these sub-objectives are summarised as follows:

(1) A multi-fidelity RBDO framework for composite structures, integrated with the use of surrogate modelling, has been developed. This was introduced in Chapter 4. The developed multi-fidelity method using ANNs provides acceptable accuracy and significant computational efficiency for carrying out the RBDO process of composite structures. First, the accuracy of the multi-fidelity models was demonstrated by comparing the conventional high-fidelity surrogate models. Overall, the

direct multi-fidelity models, which directly call the low-fidelity FEM models during the modelling process, offer more accurate solutions for reliability analyses and the RBDO process than the indirect multi-fidelity models, which call upon a surrogate model of the LFM. These multi-fidelity models using 11 high-fidelity FEM models proved to be much more accurate than a surrogate model using the same number of high-fidelity FEM models used in the multi-fidelity models. In particular, their accuracy was assured to be similar to that of a surrogate model using four times more high-fidelity FEM models. The multi-fidelity models proved to be utilisable with different reliability methods, such as MCS, FORM, and SORM. Although FORM and SORM require the first-order and second-order gradients of the response surfaces, the direct multi-fidelity models provided acceptable accuracy in estimating the probability of failure of the structural system. These multi-fidelity models also proved to be suitable for conducting the RBDO process since they correctly identified optimal solutions which are nearly identical to those obtained using the traditional high-fidelity surrogate model. The Pareto fronts between the multi-fidelity and high-fidelity surrogate models were almost similar across the solution space. The mean values and standard deviations of the selected optimal solution based on the equal mass showed a high level of accuracy between the two different models. Second, the computational efficiency was evaluated by comparing the total number of simulations required by both FEM and surrogate models for the reliability analyses and RBDO process. Notable computational time savings were achieved through the use of multi-fidelity models. All computational costs were normalised by the computation time for the MCS of the surrogate model using 100 HFMs. The direct multi-fidelity models were slightly more computationally expensive than the indirect multi-fidelity models since they directly use the low-fidelity FEM models to achieve a higher accuracy. When utilising the multi-fidelity surrogate models, the computational costs among MCS, the FORM, and the SORM were assured to be comparable due to the inherent computational benefits of the surrogate model. Given that the computational time of the surrogate model using only LFM is similar to those of the multi-fidelity models, it should be highlighted that the developed multi-fidelity probabilistic optimisation framework provides significant computational efficiency as well as high accuracy, which are nearly identical to those provided by traditional high-fidelity surrogate models.

(2) A multi-fidelity modelling formulation covering different design spaces between the HFM and the LFM has been developed. This was introduced in Chapter 5. Overall, the developed formulation allows the HFM and the LFM to have a different number of design variables, thereby offering greater computational efficiency compared to traditional multi-fidelity methods. These multi-fidelity methods, including the method presented in Chapter 4, require an equal number of FEM simulations between the HFM and the LFM. The main drawback of these methods is that they necessitate additional high-fidelity FEM simulations, associated with an increased number of design variables in the HFM. Ideally, many low-fidelity simulations and fewer high-fidelity simulations bring more computational benefits to probabilistic optimisation. The developed multi-fidelity formulation enables the HFM to cover the design spaces of a few selected design variables. At the same time, other design variables not chosen in the HFM are included in the LFM to explore the entire design space. This idea was achieved by incorporating multi-level optimisation into the multi-fidelity formulation. In this formulation, the HFM aims to find accurate optimal solutions for the selected design variables during the probabilistic optimisation process. In contrast, the LFM aims to allow engineers to explore the solution spaces of design variables not included in the HFM. It was found that the HFM discovers the optimal solutions in the high-fidelity design space at the end of each optimisation level. These optimal solutions were used to correct the multi-fidelity models before proceeding to the next level. The optimal solutions obtained by the LFM were used by the initial starting points of the next level to efficiently find global solutions. In conclusion, the multi-fidelity formulation proved to be a more computationally efficient alternative for constructing an acceptable multi-fidelity model while preserving a high level of accuracy, similar to that of traditional surrogate models.

(3) A multi-fidelity RDO framework for composite structures under the nonlinear post-buckling regime with the HFM possessing fewer design variables than the LFM has been developed. This was introduced in Chapter 5. The developed multi-fidelity formulation was incorporated into the RDO framework of composite structures. The framework was demonstrated by two engineering examples of a mono-stiffened stringer composite panel under the nonlinear post-buckling regime: deterministic optimisation (DO) and RDO. It was found that the difference in solution accuracy between the multi-fidelity model and

the high-fidelity FEM model for the DO process was less than 1%. The computational time savings obtained by the multi-fidelity model was 75% in comparison with the DO using the high-fidelity FEM model directly. The multi-fidelity methods delivered more computational gains to the RDO process while considering the design uncertainties of geometric parameters. Due to a significant number of simulations for the RDO process, the high-fidelity FEM model could not be directly used. Still, the traditional high-fidelity surrogate model was used to evaluate the efficiency and accuracy of the developed multi-fidelity methods. It was found that the objective functions obtained by the multi-fidelity model were improved as the optimisation level progressed. The difference in the optimal solutions, given an equal mass between the two models, was less than 1%, while the optimal solution using the multi-fidelity model was more robust than the high-fidelity surrogate model. It was found that more economic computation was achieved using the developed multi-fidelity model. The multi-fidelity model required both 40 high-fidelity design points and 210 low-fidelity design points. In contrast, the conventional surrogate model requested 200 high-fidelity design points. It was found that around 70% of computational time savings were obtained using this multi-level multi-fidelity method. These examples indicated the significant potential of the developed multi-fidelity method for solving large-scale composite design problems, considering design uncertainties, when compared with other multi-fidelity methods.

(4) A multi-fidelity modelling formulation utilising different sampling levels between the HFM and the LFM while considering nonlinear correlations between them has been developed. This was introduced in Chapter 6. The formulation aimed to deliver more computational gains as well as more accurate solutions. This approach was able to improve the multi-level multi-fidelity method presented in Chapter 5 by utilising the high-fidelity information about the selected design variables from the HFM. The method was capable of capturing the response surfaces of structural behaviours that show a simple linear trend between different fidelity models. The formulation presented in this chapter was motivated by using different sampling levels to collect high-fidelity information from the entire design space while collecting as much low-fidelity information as possible. The sampling strategy was established to collect high-fidelity information using dense sampling and sparse sampling. It was found that this sampling strategy allows the multi-fidelity model to

require a smaller number of high-fidelity FEM simulations embracing a part of the entire design space. At the same time, it was found that the high-fidelity training dataset offers sparse information on other design spaces without the need for additional high-fidelity FEM simulations. A nonlinear fusion GP was used to consider the nonlinear correlation between the HFM and the LFM. Compared to the use of conventional surrogate modelling methods, the use of the nonlinear fusion GP enabled the construction of a more accurate multi-fidelity model, capturing a complex trend between different fidelity models. Both the sampling scheme and the nonlinear fusion GP were integrated with multi-level optimisation, thereby enhancing computational efficiency for large-scale design problems. In conclusion, it has been demonstrated that this novel multi-fidelity modelling formulation can efficiently and accurately conduct the probabilistic optimisation of composite structures.

(5) A multi-fidelity probabilistic optimisation framework for composite structures subjected to thermo-mechanical loading has been developed. This was introduced in Chapter 6. The developed multi-fidelity formulation using the nonlinear fusion GP was incorporated into the RBDO process. The established framework was demonstrated using an engineering example of a mono-stiffened stringer panel under thermo-mechanical loading. The composite panel was subjected to mechanical loading and thermal loading. The design uncertainties of four geometric parameters and six material properties were considered as input parameters for constructing the multi-fidelity model. It was found that the multi-fidelity model for each level of the RBDO process scrutinises a part of the design spaces while providing sparse high-fidelity information to the LFM. This enabled the multi-fidelity model to be updated using the optimal solutions at the end of each level, thus preparing it to explore different design spaces for the following level. The Pareto front of the multi-fidelity model was compared with the conventional surrogate model using 150 high-fidelity FEM simulations. It was found that the solution spaces are more thoroughly examined as the optimisation level goes up. This offered more reliable and better solutions compared with those of the previous optimisation levels since the design space to be explored by the multi-fidelity model narrowed down. The statistical characteristics of the optimal solution obtained by the multi-fidelity model were nearly identical to those of the optimal solution from the high-fidelity surrogate model.

The multi-fidelity model proved to be more accurate than the surrogate model using 40 high-fidelity FEM simulations, which equates to using both 20 HFMs and 70 LFMs required for constructing the multi-fidelity model. In terms of computational efficiency, it is not surprising that the multi-fidelity model accomplished about a 75% reduction compared to the surrogate model using 150 high-fidelity FEM simulations. It was found that the multi-fidelity model delivers around 50% computational time savings compared to the traditional multi-fidelity method [56] using both 60 HFMs and 60 LFMs. This multi-fidelity model proved to be approximately 30% more computationally efficient than the multi-level multi-fidelity method presented in Chapter 5, which required both 20 HFMs and 110 LFMs. Given that the computational cost of low-fidelity FEM simulations could lead to a serious challenge when it comes to large-scale design problems, the gains obtained from a smaller number of low-fidelity FEM models can be significant.

References

[1] Jones RM. *Mechanics of Composite Materials.* Taylor and Francis, Milton Park, Abingdon-on-Thames, Oxfordshire, U.K; 1997.

[2] Baker JW, Schubert M, Faber MH. On the assessment of robustness. *Structural Safety.* 2008;30:253–267. doi:10.1016/j.strusafe.2006.11.004.

[3] Taguchi G. *Introduction to Quality Engineering: Designing Quality into Products and Processes.* Quality Resources, Cleveland, Ohio, United States; 1986.

[4] Bacarreza O, Aliabadi MH, Apicella A. Robust design and optimization of composite stiffened panels in post-buckling. *Structural and Multidisciplinary Optimization.* 2015;51:409–422. doi:10.1007/s00158-014-1136-5.

[5] Long MW and Narcosi JD. Probabilistic design methodology for composite aircraft structure. Final Report. U.S. Department of Transportation; 1999. Available from: http://www.tc.faa.gov/its/worldpac/techrpt/ar99-2.pdf.

[6] Bouckaert S, Pales AF, McGlade C, Remme U, Wanner B. Net-zero by 2050: a roadmap for the global energy sector. Technical Report. *International Energy Agency*; 2021. Available from: https://www.iea.org/reports/net-ze ro-by-2050.

[7] Rackwitz R. Reliability analysis — a review and some perspectives. *Structural Safety.* 2001;23:365–95. Available from: https://doi.org/10.1016/ S0167-4730(02)00009-7.

[8] Choi S-K, Grandhi R, Canfield RA. *Reliability-based Structural Design.* Springer, Berlin, Germany; 2006. doi:10.1007/978-1-84628-445-8.

[9] Beck AT, Melchers RE. *Structural Reliability Analysis and Prediction.* John Wiley and Sons, Hoboken, New Jersey, U.S.; 2018. doi:10.1002/ 9781119266105.

[10] Wang GG, Shan S. Review of metamodeling techniques in support of engineering design optimization. *Journal of Mechanical Design.* 2007;129: 370–380. doi:10.1115/1.2429697.

[11] Forrester AI, Sobester A, Keane AJ. *Engineering Design via Surrogate Modelling: A Practical Guide*. John Wiley and Sons, Hoboken, New Jersey, U.S.; 2008. doi:10.1002/9780470770801.

[12] Zhao Y, Jiang C, Vega MA, Todd MD, Hu Z. Surrogate modeling of nonlinear dynamic systems: a comparative study. *ASME Journal of Computing and Information Science in Engineering*. 2023. doi:10.1115/1.4054039.

[13] Peherstorfer B, Willcox K, Gunzburger M. Survey of multi-fidelity methods in uncertainty propagation, inference, and optimization. *SIAM Review*. 2018;60:550–591. doi:10.1137/16M1082469.

[14] Das TK, Ghosh P, Das NC. Preparation, development, outcomes, and application versatility of carbon fiber-based polymer composites: a review. *Advanced Composites and Hybrid Materials*. 2019;2:214–233. doi:10.1007/s42114-018-0072-z.

[15] López C, Bacarreza O, Baldomir A, Hernández S, Aliabadi MHF, Reliability-based design optimization of composite stiffened panels in post-buckling regime. *Structural and Multidisciplinary Optimization*. 2017;55:1121–1141. doi:10.1007/s00158-016-1568-1.

[16] Farokhi H, Bacarreza O, Aliabadi MHF. Probabilistic optimisation of mono-stringer composite stiffened panels in post-buckling regime. *Structural and Multidisciplinary Optimization*. 2020;62:1395–1417. doi:10.1007/s00158-020-02565-9.

[17] Liu X, Qin J, Zhao K, Featherston CA, Kennedy D, Jing Y, Yang G. Design optimization of laminated composite structures using artificial neural network and genetic algorithm. *Composite Structures*. 2023;305. doi:10.1016/j.compstruct.2022.116500.

[18] Álvarez JG, Bisagni C. Closed-form solutions for thermomechanical buckling of orthotropic composite plates. *Composite Structures*. 2020;233:111622. doi:10.1016/j.compstruct.2019.111622.

[19] Lee JJ, Choi S. Thermal buckling and postbuckling analysis of a laminated composite beam with embedded SMA actuators. *Composite Structures*. 1999;47:695–703. doi:10.1016/S0263-8223(00)00038-6.

[20] Yoo K-K, Kim J-H. Optimal design of smart skin structures for thermomechanical buckling and vibration using a genetic algorithm. *Journal of Thermal Stresses*. 2011;34. doi:10.1080/01495739.2011.601261.

[21] Guerrero JM, Sasikumar A, Llobet J, Costa J. Experimental and virtual testing of a composite-aluminium aircraft wingbox under thermal loading. *Aerospace Science and Technology*. 2023;128. doi:10.1016/j.ast.2023.108329.

[22] Yao W, Chen X, Luo W, Tooren MV, Guo J. Review of uncertainty-based multidisciplinary design optimization methods for aerospace vehicles. *Progress in Aerospace Sciences*. 2011;47:450–479. doi:10.1016/j.paerosci.2011.05.001.

[23] Sharma A, Mukhopadhyay T, Rangappa SM. Advances in computational intelligence of polymer composite materials: machine learning assisted modeling, analysis and design. *Archives of Computational Methods in Engineering*. 2022;29. doi:10.1007/s11831-021-09700-9.

[24] Sbaraglia F, Farokhi H, Aliabadi MHF. Robust and reliability-based design optimization of a composite floor beam. *Key Engineering Materials.* 2018;774:486–491. doi:10.4028/www.scientific.net/KEM.774.486.

[25] Acar E, Bayrak G, Jung Y. Modeling, analysis, and optimization under uncertainties: a review. *Structural and Multidisciplinary Optimization.* 2021;64. doi:10.1007/s00158-021-03026-7.

[26] Viana FA, Simpson TW, Balabanov V, Toropov V. Metamodeling in multidisciplinary design optimization: how far have we really come? *AIAA Journal.* 2014;52:670–690. doi:10.2514/1.J052375.

[27] Babu SS, Mourad AHI, Harib KH, Vijayavenkataraman S. Recent developments in the application of machine-learning towards accelerated predictive multiscale design and additive manufacturing. *Virtual and Physical Prototyping.* 2022;18. doi:10.1080/17452759.2022.2141653.

[28] Parnianifard A, Azfanizam AS, Ariffin MK, Ismail MI, Ebrahim NA. Recent developments in metamodel based robust black-box simulation optimization: an overview. *Decision Science Letters.* 2019;8:17–44. doi: 10.5267/j.dsl.2018.5.004.

[29] Fernández-Godino MG, Park C, Kim N-H, Haftka RT. Review of multifidelity models. 2016. arXiv preprint arXiv:1609.07196v3. doi:10.1016/j.jcp. 2015.01.034.

[30] Toal DJJ. Applications of multi-fidelity multi-output kriging to engineering design optimization. *Structural and Multidisciplinary Optimization.* 2023;66. doi:10.1007/s00158-023-03567-z.

[31] do Vale JL, Sohst M, Crawford C, Suleman A, Potter G, Banerjee S. On the multi-fidelity approach in surrogate-based multidisciplinary design optimisation of high-aspect-ratio wing aircraft. *The Aeronautical Journal.* 2023;127. doi:10.1017/aer.2022.49.

[32] Yoo K, Bacarreza O, Aliabadi MHF. A novel multi-fidelity modelling-based framework for reliability-based design optimisation of composite structures. *Engineering with Computers.* 2020. doi:10.1007/s00366-020-01084-x.

[33] Andradottir S. A review of simulation optimization techniques. *In: 1998 Winter Simulation Conference. Proceedings (Cat. No.98CH36274)*, Vol. 1; 1998. p. 151–158. doi:10.1109/WSC.1998.744910.

[34] Haftka RT, Gürdal Z. *Elements of structural optimization.* Dordrecht: Springer; 1992. doi:https://doi.org/10.1007/978-94-011-2550-5.

[35] Marler RT, Arora JS. Survey of multi-objective optimization methods for engineering. *Structural and Multidisciplinary Optimization.* 2004;26:369–395. doi:10.1007/s00158-003-0368-6.

[36] Boyd S, Vandenberghe L. *Convex optimization.* Cambridge University Press, Cambridge, U.K; 2004. doi:10.1017/CBO9780511 804441.

[37] Vanderplaats GN. *Multidiscipline Design Optimization.* Vanderplaats Research and Development, Michigan, United States; 2007.

[38] Koziel S, Yang X-S. *Computational Optimization: Methods and Algorithms*, Vol. 356. Berlin: Springer; 2011. doi:10.1007/978-3-642-20859-1.

[39] Arora RK. *Optimization: Algorithms and applications*. Chapman and Hall/CRC, Cambridge, U.K; 2015. doi:10.1017/cbo9780511974076.010.

[40] Qureshi MT, and Yoo K. Assessment of algorithms for the probabilistic optimization of composite panels. *AIP Conference Proceedings*. New York, United States, 2020.

[41] Snyman J. *Practical Mathematical Optimization: An Introduction to Basic Optimization Theory and Classical and New Gradient-based Algorithms*. Applied Optimization. Springer US, New York, United States; 2005. Available from: https://books. google.co.uk/books?id=Md_XvAi1BvcC.

[42] Vervenne K. *Gradient-based Approximate Design Optimization*. Delft University Press, Delft, Netherlands; 2005. Available from: https://books.goog le.co.uk/books?id=IOovPQAACAAJ.

[43] Lewis RM, Torczon V, Trosset MW. Direct search methods: then and now. *Journal of Computational and Applied Mathematics*. 2000;124. doi:10.1016/ S0377-0427(00)00423-4.

[44] Zitzler E, Lothar T, Kalyanmoy D. Comparison of multiobjective evolutionary algorithms: empirical results. *Evolutionary Computation*. 2000;8:173–195. doi:10.1016/S0377-0427(00)00423-4.

[45] Deb K. *Multiobjective Optimization Using Evolutionary Algorithms*. Wiley, New Jersey, United States; 2001.

[46] Coello CA. An updated survey of ga-based multiobjective optimization techniques. *ACM Computing Surveys*. 2000;32:109–143. doi:10.1145/ 358923.358929.

[47] Kim C, Kwon YW. Reliability-based design considering prediction interval estimation to optimize composite patches. *Mechanics Based Design of Structures and Machines*. 2022;52. doi:10.1080/15397734.2022.2159836.

[48] Bektas E, Broermann K, Pecanac G, Rzepka S, Silber C, Wunderle B. Robust design optimization: on methodology and short review. In: *2017 18th International Conference on Thermal, Mechanical and Multi-Physics Simulation and Experiments in Microelectronics and Microsystems, EuroSimE 2017*; 2017. pp. 1–7. doi:10.1109/EuroSimE.2017.7926290.

[49] Hozic D, Thore C-J, Cameron C, Loukil M. Deterministic-based robust design optimization of composite structures under material uncertainty. *Composite Structures*. 2023;322. doi:j.compstruct.2023.117336.

[50] Morse L, Khodaei ZS, Aliabadi MH. Multi-fidelity modeling-based structural reliability analysis with the boundary element method. *Journal of Multiscale Modelling*. 2017;8:1740001. doi:10.1142/S1756973717400017.

[51] Yoo K, Bacarreza O, Aliabadi MH. Multi-fidelity robust design optimisation for composite structures based on low-fidelity models using successive high-fidelity corrections. *Composite Structures*. 2021;259:113477. doi:10.1016/j. compstruct.2020.113477.

[52] Mukhopadhyay T, Naskar S, Dey NS. On machine learning assisted data-driven bridging of FSDT and HOZT for high-fidelity uncertainty quantification of laminated composite and sandwich plates. *Composite Structures*. 2023;304. doi:j.compstruct.2022.116276.

[53] Schuëller GI, Jensen HA. Computational methods in optimization considering uncertainties — an overview. *Computer Methods in Applied Mechanics and Engineering.* 2008;198:2–13. doi:10.1016/j.cma.2008.05.004.

[54] Youn BD, Choi KK. Selecting probabilistic approaches for reliability-based design optimization. *AIAA Journal.* 2008;42:124–131. doi:10.2514/1.9036.

[55] Haldar A, Mahadevan S. *Probability, Reliability, and Statistical Methods in Engineering Design.* Wiley, New Jersey, United States; 1999.

[56] Vitali R, Haftka RT, Sankar BV. Multi-fidelity design of stiffened composite panel with a crack. *Structural and Multidisciplinary Optimization.* 2002;23:347–356. doi:10.1007/s00158-002-0195-1.

[57] Park C, Haftka RT, Kim NH. Remarks on multi-fidelity surrogates. *Structural and Multidisciplinary Optimization.* 2017;55:1029–1050. doi:10.1007/s00158-016-1550-y.

[58] Burhenne S, Jacob D, Henze GP. Sampling-based on sobol sequence for monte carlo techniques applied to building simulation. In: *Proceedings of Building Simulation 2011: 12th Conference of International Building Performance Simulation Association*; 2011. pp. 1816–1823.

[59] Hassoun MH. *Fundamentals of Artificial Neural Networks.* MIT Press, Massachusetts, United States; 1995. doi:10.1109/tnn.1996.501738.

[60] Buhmann MD. *Radial Basis Functions.* Cambridge University Press, Cambridge, U.K; 2003. doi:10.1017/CBO9780511543241.

[61] Fang H, Horstemeyer MF. Global response approximation with radial basis functions. *Engineering Optimization.* 2006;38. doi:10.1080/03052150500422294.

[62] Magdon-Ismail MM, Lin H-T, Abu-Mostafa YS. *Learning From Data.* AMLBook, United States; 2012.

[63] Rasmussen CE, Williams CKI, Bach F. *Gaussian Processes for Machine Learning,* 2nd edn. MIT Press, Massachusetts, United States; 2006. Available from: http://www.gaussianprocess.org/gpml/chapters/RW.pdf.

[64] Mack Y, Goel T, Shyy W, Haftka R. Surrogate model-based optimization framework: a case study in aerospace design. Springer, Berlin, Germany; 2007. doi:10.1007/978-3-540-49774-5_14.

[65] Eason J, Cremaschi S. Adaptive sequential sampling for surrogate model generation with artificial neural networks. *Computers and Chemical Engineering.* 2014;68:220–232. doi:10.1016/j.compchemeng.2014.05.021.

[66] SIMULIA 3ds. *Abaqus Documentation.* 3ds; 2019.

[67] SIMULIA 3ds. *Isight Documentation.* Simulia; 2019. http://www.3ds.com/fileadmin/PRODUCTS/SIMULIA/PDF/whitepapers/SIMULIA-Isight-Design-Optimization-Methodologies.pdf.

[68] Lam R. Surrogate modeling-based on statistical techniques for multi-fidelity optimization [PhD thesis]. Massachusetts Institute of Technology; 2014. Available from: http://hdl.handle.net/1721.1/90673.

[69] Gano SE, Renaud JE, Sanders B, Hybrid variable fidelity optimization by using a kriging-based scaling function. *AIAA Journal.* 2008;43:2422–2433. doi:10.2514/1.12466.

[70] Bakr MH, Bandler JW, Madsen K, Søndergaard J. An introduction to the space mapping technique. *Optimization and Engineering.* 2001;2:369–384. doi:10.1023/A:1016086220943.

[71] Bakr MH, Bandler J, Madsen K, Søndergaard J. Review of the space mapping approach to engineering optimization and modeling. *Optimization and Engineering.* 2000;1:241–276. doi:10.1023/A:1010000106286.

[72] Forrester AI, Sóbester A, Keane AJ. Multi-fidelity optimization via surrogate modelling. *Proceedings of the Royal Society A: Mathematical, Physical and Engineering Sciences.* 2007;463:3251–3269. doi:10.1098/rspa. 2007.1900.

[73] Kennedy MC, O'Hagan A. Predicting the output from a complex computer code when fast approximations are available. *Biometrika.* 2000;87:1–13. doi: 10.1093/biomet/87.1.1.

[74] Hassanien S, Kainat M, Adeeb S, Langer D. On the use of surrogate models in reliability-based analysis of dented pipes. In: *Proceedings of the 2016 11th International Pipeline Conference.* ASME; 2016. pp. 1–9. doi:10.1115/ IPC2016-64470.

[75] Lee S, Kim I-G, Cho W, Shul C. Advanced probabilistic design and reliability-based design optimization for composite sandwich structure. *Advanced Composite Materials.* 2014;23:3–16. doi:10.1080/09243046.2013. 862381.

[76] Scarth C, Sartor PN, Cooper JE, Weaver PM, Silva GHC. Robust and reliability-based aeroelastic design of composite plate wings. *AIAA Journal.* 2017;55:3539–3552. doi:10.2514/1.J055829.

[77] Jin R, Chen W, Sudjianto A. On sequential sampling for global meta-modeling in engineering design. In: *Proceedings of DETC'02: ASME 2002 Design Engineering Technical Conferences And Computers and Information in Engineering Conference.* ASME; 2002. DETC2002/DAC–34092. doi: 10.1115/DETC2002/DAC-34092.

[78] Alexandrov NM, Lewis RM, Gumbert CR, Green LL, Newman PA. Approximation and model management in aerodynamic optimization with variable-fidelity models. *Journal of Aircraft.* 2001;38:1093–1101. doi:10.2514/2.2877.

[79] Goldfeld Y, Vervenne K, Arbocz J, van Keulen F. Multi-fidelity optimization of laminated conical shells for buckling. *Structural and Multidisciplinary Optimization.* 2005;30:128–141. doi:10.1016/j.tws.2004.07.003.

[80] Causon D, Mingham C. *Introductory Finity Difference Methods for PDEs.* Bookboon; 2010.

[81] Lee CY, Kim JH. Thermo-mechanical characteristics and stability boundaries of antenna structures in supersonic flows. *Composite Structures.* 2013;97:363–369. doi:10.1016/j.compstruct.2012.09.048.

[82] Zhou XY, Ruan X, Gosling PD. Robust design optimization of variable angle tow composite plates for maximum buckling load in the presence of uncertainties. *Composite Structures.* 2019;223. doi:10.1016/j.compstruct. 2019.110985.

[83] António CC, Hoffbauer LN. Bi-level dominance ga for minimum weight and maximum feasibility robustness of composite structures. *Composite Structures.* 2016;135:83–95. doi:10.1016/j.compstruct.2015.09.019.

[84] Keane AJ. Cokriging for robust design optimization. *AIAA Journal.* 2012;50:2351–2364. doi:10.2514/1.j051391.

[85] Vuruskan A, Hosder S. Impact of turbulence models and shape parameterization on robust aerodynamic shape optimization. *Journal of Aircraft.* 2019;56:1099–1115. doi:10.2514/1.C035039.

[86] Dodson M, Parks GT. Robust aerodynamic design optimization using polynomial chaos. *Journal of Aircraft.* 2009;46:635–646. doi:10.2514/1.39419.

[87] Ong YS, Nair PB, Keane AJ. Evolutionary optimization of computationally expensive problems via surrogate modeling. *AIAA Journal.* 2003;41:687–696. doi:10.2514/2.1999.

[88] Mohammed A, Harris I, Govindan K. A hybrid mcdm-fmoo approach for sustainable supplier selection and order allocation. *International Journal of Production Economics.* 2019;217:171–184. doi:10.1016/j.ijpe.2019.02.003.

[89] Gu X, Sun G, Li G, Mao L, Li Q. A comparative study on multiobjective reliable and robust optimization for crashworthiness design of vehicle structure. *Structural and Multidisciplinary Optimization.* 2013;48:669–684. doi:10.1007/s00158-013-0921-x.

[90] Zhou Q, Wang Y, Choi SK, Jiang P, Shao X, Hu J, Shu L. A robust optimization approach based on multi-fidelity metamodel. *Structural and Multidisciplinary Optimization.* 2018;57:775–797. doi:10.1007/s00158-017-1783-4.

[91] Robinson TD, Eldred MS, Willcox KE, Haimes R. Surrogate-based optimization using multi-fidelity models with variable parameterization and corrected space mapping. *AIAA Journal.* 2008;46:2814–2822. doi:10.2514/1.36043.

[92] Leary SJ, Bhaskar A, Keane AJ. A constraint mapping approach to the structural optimization of an expensive model using surrogates. *Optimization and Engineering.* 2001;2. doi:10.1023/A:1016038305014.

[93] Madsen K, Søndergaard J. Convergence of hybrid space mapping algorithms. *Optimization and Engineering.* 2004;5:145–156. doi:10.1023/b:opte.0000033372.34626.49.

[94] Sinha A, Malo P, Deb K. A review on bi-level optimization: from classical to evolutionary approaches and applications. *IEEE Transactions on Evolutionary Computation.* 2018;22:276–295. doi:10.1109/TEVC.2017.2712906.

[95] Choi S, Alonso JJ, Kroo IM. Two-level multi-fidelity design optimization studies for supersonic jets. *Journal of Aircraft.* 2009;46:776–790. doi:10.2514/1.34362.

[96] Liu B, Haftka RT, Akgün MA. Two-level composite wing structural optimization using response surfaces. *Structural and Multidisciplinary Optimization.* 2000;20:87–96. doi:10.1007/s001580050140.

[97] Stevens K, Ricci R, Davies G. Buckling and postbuckling of composite structures. *Composites.* 1995;26. doi:10.1016/0010-4361(95)91382-F.

[98] Bisagni C. Numerical analysis and experimental correlation of composite shell buckling and post-buckling. *Composites Part B: Engineering.* 2000;31:655–667. doi:10.1016/S1359-8368(00)00031-7.

[99] Tsai SW, Wu EM. A general theory of strength for anisotropic materials. *Journal of Composite Materials*. 1971;5. doi:10.1177/002199837100500106.

[100] Akcin Y, Karakaya S, Soykasap O. Electrical, thermal and mechanical properties of cnt treated prepreg cfrp composites. *Materials Sciences and Applications*. 2016;07:465–483. doi:10.4236/msa.2016.79041.

[101] Wang C, Xu Y, Du J. Study on the thermal buckling and post-buckling of metallic sub-stiffening structure and its optimization. *Materials and Structures/Materiaux et Constructions*. 2016;49:4867–4879. doi:10.1617/s11527-016-0830-8.

[102] Vosoughi AR, Nikoo MR. Maximum fundamental frequency and thermal buckling temperature of laminated composite plates by a new hybrid multi-objective optimization technique. *Thin-Walled Structures*. 2015;95:408–415. doi:10.1016/j.tws.2015.07.014.

[103] Meyers CA, Hyer MW. Thermal buckling and post-buckling of symmetrically laminated composite plates. *Journal of Thermal Stresses*. 1991;14. doi:10.1080/01495739108927083.

[104] Nawab Y, Jacquemin F, Casari P, Boyard N, Borjon-Piron Y, Sobotka V. Study of variation of thermal expansion coefficients in carbon/epoxy laminated composite plates. *Composites Part B: Engineering*. 2013;50. doi:10.1016/j.compositesb.2013.02.002.

[105] Chen X, Qiu Z. Reliability assessment of fiber-reinforced composite laminates with correlated elastic mechanical parameters. *Composite Structures*. 2018;203:396–403. doi:10.1016/j.compstruct.2018.05.032.

[106] Sohouli A, Yildiz M, Suleman A. Efficient strategies for reliability-based design optimization of variable stiffness composite structures. *Structural and Multidisciplinary Optimization*. 2017;1–16. doi:10.1007/s00158-017-1771-8.

[107] das Neves Carneiro G, Antonio CC. A RBRDO approach based on structural robustness and imposed reliability level. *Structural and Multidisciplinary Optimization*. 2018;57:2411–2429. doi:10.1007/s00158-017-1870-6.

[108] Perdikaris P, Venturi D, Royset JO, Karniadakis GE. Multi-fidelity modelling via recursive co-kriging and gaussian-markov random fields. *Proceedings of the Royal Society A: Mathematical, Physical and Engineering Sciences*. 2015;471. doi:10.1098/rspa.2015.0018.

[109] Perdikaris P, Raissi M, Damianou A, Lawrence ND, Karniadakis GE. Nonlinear information fusion algorithms for data-efficient multi-fidelity modelling. *Proceedings of the Royal Society A: Mathematical, Physical and Engineering Sciences*. 2017;473. doi:10.1098/rspa.2016.0751.

[110] Yoo K, Bacarreza O, Aliabadi MF. Multi-fidelity probabilistic optimisation of composite structures under thermomechanical loading using gaussian processes. *Computers and Structures*. 2021;257. doi:10.1016/j.compstruc.2021.106655.

[111] Gratiet LL. Recursive co-kriging model for design of computer experiments with multiple levels of fidelity with an application to hydrodynamic; 2013. arXiv:1210.0686 [math.ST].

[112] Bouhlel MA, Bartoli N, Otsmane A, Morlier J. An improved approach for estimating the hyperparameters of the Kriging model for high-dimensional problems through the partial least squares method. *Mathematical Problems in Engineering*. 2016;6723410. doi:10.1155/2016/6723410.

Index

Computational and Experimental Methods in Structures

(Continued from page ii)

Computational and Experimental Methods in Structures

(Continued from page ii)